A FIELD GUIDE AN͞D INTRODUCTION TO
THE GEOLOGY A͞ND CHEMISTRY OF

ROCKS AND
MINERALS

Original Title: MINERALS OF THE WORLD

by CHARLES A. SORRELL

Professor of Ceramic Engineering
School of Mines and Metallurgy
University of Missouri, Rolla

Illustrated by GEORGE F. SANDSTRÖM

Original Project Editor: HERBERT S. ZIM

Cinnabar
Piute Co., Utah

 GOLDEN PRESS • NEW YORK
Western Publishing Company, Inc.
Racine, Wisconsin

FOREWORD

Of all the objects of nature, minerals are among the most readily available for study and, because of their immense variety and intrinsic beauty, have always been items of interest and curiosity. This book was written to fill the gap between available popular books and the typical college textbook; it emphasizes chemical relationships and crystal structures with some introduction to rocks and their geologic relationship to major rock types. The hundreds of full color illustrations provide the mineral enthusiast with a reference catalog for identification not available in any other book.

Without the generous cooperation of many individuals, the illustrations for this guide could not have been prepared. We are therefore indebted to Dr. James E. Bever of Miami University, Oxford, Ohio and to Dr. John C. Butler and Dr. George Distler, who, during their graduate studies, assisted with the selection of specimens from the Miami University geology collection. Ward's Natural Science Establishment, Rochester, New York, was helpful in providing several specimens on loan, and Dr. Kerry Grant, University of Missouri-Rolla generously made available their geology collection.

Generous help, advice and full cooperation, especially in obtaining and identifying needed specimens, was given the artist by the staff of the Geology Department of Bryn Mawr College, Bryn Mawr, Pennsylvania, under the chairmanship of Associate Professor, Lucien B. Platt. Special mention and gratitude is expressed to Professor Emeritus of Geology and Research Associate, Edward H. Watson; Assistant Professor of Geology, William A. Crawford and Maria Luisa Crawford (Curator); and Assistant Curator, Harold W. Ardnt, who unselfishly gave much of his time in locating needed specimens. Recognition is also due Edward McHale (deceased). Specimens supplied to the artist by Bryn Mawr College came from the George S. Vaux and Theodore Rand collections.

The artist is also grateful for the helpful advice and cooperation, received at the beginning of this project, from Paul Seel and Walter Groff (deceased), both amateur mineralogists.

ROCKS AND MINERALS is another addition to the Golden Field Guide series, which presently includes BIRDS OF NORTH AMERICA, SEASHELLS OF NORTH AMERICA, and TREES OF NORTH AMERICA.

C.A.S.
G.F.S.

TABLE OF CONTENTS

Villiaumite
(p. 122)

Acanthite
Freiberg, E. Germany
(p. 94)

Stromeyerite
Broken Hills, N.S.W.,
Australia
(p. 94)

Berzelianite
with Calcite
Tisnov, Czechoslovakia
(p. 94)

MINERALOGY AS A SCIENCE

Minerals are the natural crystalline materials that form the Earth and make up most of its rocks. Though minerals have been used and metals extracted from them for all of recorded history, mineralogy as a science is relatively young. Serious study of minerals began in the 1800's, after the development of the petrographic microscope (for studying rocks) and the reflecting goniometer (for accurately measuring angles between faces of a mineral's crystals). During that century most of the minerals known today were described, optically studied, and chemically analyzed. Largely from these studies grew the schemes used today to classify minerals. The internal crystal structure of minerals, however, could only be guessed from their external symmetry and optical properties.

Wilhelm Roentgen's discovery of X-rays in 1895 provided mineralogists with the tool they needed to study crystal structures, but it was not until 1912 that Max von Laue and his assistants proved that when X-rays are scattered by a crystal their behavior gives clues to the internal arrangement. Since then, the structures of all important mineral groups have been analyzed. By correlating this structural knowledge with physical, chemical, electrical, thermal, and mechanical properties, mineralogists have gained an understanding of the forces that hold crystalline matter together. This understanding has in turn enabled scientists to synthesize crystalline compounds with properties to fill special needs. These compounds have been used in the manufacture of high-temperature ceramics, electrical insulators, transistors, and many other items.

Knowledge of the behavior of crystals at high temperatures and pressures has been applied to research on the formation of mountains, the eruption of volcanoes, and other geologic processes. The forces that cause these activities become more understandable when analyzed in terms of structural changes in the mineral components of rock.

Rocks, metals, concrete, bricks, plaster, paint pigments, paper, rubber, and ceramics all contain mineral or synthetic crystals. In fact, almost all solids except glass and organic materials are crystalline. This is why knowledge of the structure and behavior of crystals is important in nearly all industrial and technical endeavors. Even organic materials form crystals when isolated in a pure state. By studying these crystals, biologists and medical researchers have learned much about life processes and heredity.

Unquestionably mineralogy is a subject of widespread importance in all of science. Consequently, persons trained in mineralogy and crystallography may be found doing work in the parent science, geology, or may be engaged in research in metallurgical, ceramic, or polymer materials, in solid state physics or chemistry, or in the biological sciences. Interdisciplinary cooperation among scientists has led to many important discoveries.

MINERALS IN ROCKS

Minerals are the constituents of rocks, which make up the entire inorganic, solid portion of the earth. Mineral formation and rock formation are, in fact, one process. To know minerals, therefore, it is important to know rocks. A single mineral may form a rock, but usually rocks are cohesive aggregates of two or more minerals. Depending on how they were formed, rocks are divided into three types: igneous, metamorphic, and sedimentary.

IGNEOUS ROCKS are formed by the cooling and hardening of magma, a complex molten material that originates within the earth. Some important types of igneous rocks are shown in the illustration on the facing page. The major mineral constituents of acid, intermediate, and basic rocks shown provide the basis for the classification given on page 9.

IGNEOUS MINERALS important in the formation of igneous rocks are relatively few in number. This is because the magma from which the minerals crystallize is rich only in certain elements: silicon, oxygen, aluminum, sodium, potassium, calcium, iron, and magnesium. These are the elements that combine and form the silicate minerals (pp. 154-227). A limited number of the silicates —the olivines, pyroxenes, amphiboles, micas, feldspars, and quartz—account for over 90 percent of all igneous rocks.

As magma cools, minerals crystallize at different temperatures. Olivine and calcium feldspar form at high temperatures and may separate early from the melt. Other minerals solidify as the temperature falls (see Bowen's Reaction Series, pp. 82-83). The last to crystallize are potassium feldspar, muscovite mica, and quartz, the major constituents of granite. Finally, water in the magma, carrying valuable metals and sulfur in solution, moves outward through fractures in the surrounding rock and deposits sulfides in veins. The water is also important because it affects the temperature at which crystallization occurs and the types of minerals formed during cooling.

INTRUSIVE IGNEOUS ROCKS, also called plutonic rocks, crystallize from magma that cools and hardens within the earth. Surrounded by pre-existing rock, called country rock, the magma cools slowly. As a result, these rocks are coarse-grained.

Central cores of major mountain ranges consist of large masses of plutonic rock, generally granite, intruded as a part of the mountain-building process. When exposed by erosion, these cores, called batholiths, may occupy millions of square miles of surface area. Offshoots of batholiths bear different names, such as laccoliths and sills, depending on their size and their relationship to the country rock. The term abyssal is commonly used to describe coarse-grained rocks formed at depth; hypabyssal is used to describe intrusive rocks formed near the surface.

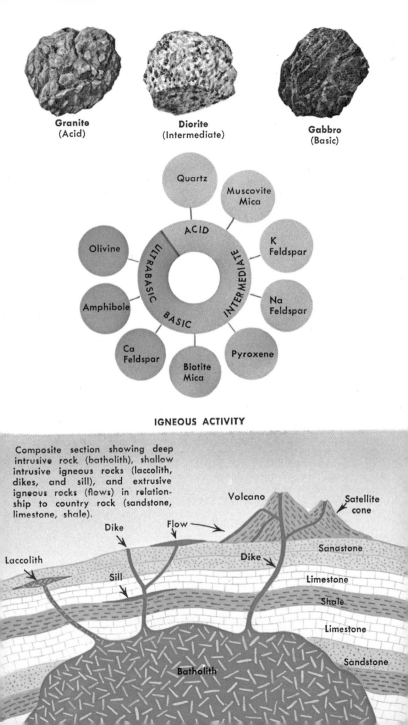

Granite
(Acid)

Diorite
(Intermediate)

Gabbro
(Basic)

Quartz

Muscovite
Mica

Olivine

K
Feldspar

ACID

ULTRABASIC

INTERMEDIATE

Amphibole

BASIC

Na
Feldspar

Ca
Feldspar

Biotite
Mica

Pyroxene

IGNEOUS ACTIVITY

Composite section showing deep intrusive rock (batholith), shallow intrusive igneous rocks (laccolith, dikes, and sill), and extrusive igneous rocks (flows) in relationship to country rock (sandstone, limestone, shale).

Volcano

Satellite
cone

Dike

Flow

Laccolith

Dike

Sill

Sanastone

Limestone

Shale

Limestone

Sandstone

Batholith

EXTRUSIVE IGNEOUS ROCKS, also called volcanic rocks, are formed at the earth's surface as a result of volcanic activity. Like batholith formation (p. 6), this activity is associated with mountain-building forces within the earth. Temperatures only a few miles beneath the earth's surface are higher than the temperatures at which most rocks would melt at the surface. The below-surface rocks remain solid, however, because of the pressure exerted by overlying rocks. If the rocks fracture—as the result of mountain-building forces, for example—the pressure may be released, and a sizable volume of rock will melt. The resulting magma will be forced through the fractures to the surface, forming a volcano.

Molten rock, or lava, will flow from the volcano and spread onto the ground. Because the lava cools and crystallizes rapidly, it is fine-grained. Material may be blown violently from the volcanic pipe as blocks, pellets, and dust, or as a liquid that hardens before it falls to the surface. These pyroclastics may fall nearby, forming part of the volcano, or may be spread great distances by winds.

CLASSIFICATION of the many and greatly different kinds of igneous rocks can provide important information as to the conditions of formation. Two obvious variables that may be used as criteria for classification are particle size, which depends largely on cooling history, and composition, both chemical and mineralogical. Because feldspars, quartz, olivines, pyroxenes, amphiboles, and micas are the important minerals in the formation of igneous rocks, they are basic to the classification of those rocks. All other minerals are nonessential (accessory).

In the simplified classification on the opposite page, rock types are separated on the basis of the type of feldspar present, the presence or absence of quartz, and, in rocks with no feldspar or quartz, the type of iron and magnesium minerals present. Rocks with crystals large enough to be seen by the eye are called phaneritic; those with crystals too small to be seen are called aphanitic. In general, phaneritic implies an intrusive origin; aphanitic, an extrusive origin. Porphyritic refers to crystals embedded in a fine-grained rock. More detailed classifications using these terms are given in geology and petrology texts.

GRANITES show evidence of being the result of either igneous or metamorphic processes. Some granites obviously have crystallized from a melt; blocks of partially assimilated country rock may be found in granite, clearly indicating that the country rock fell into a liquid magma that hardened around it. Other granites, however, bear evidence of having been formed by metamorphism (p. 10); variations in composition of pre-existing sedimentary rocks are reflected in banding preserved in the granite. The conversion of sedimentary rock to granite by metamorphism is called granitization.

Rhyolite
(Acid lava)

Andesite
(Intermediate lava)

Basalt
(Basic lava)

Tuff
(Cemented ash)

Cinder
(Pyroclastic debris)

Obsidian
(Volcanic glass))

EXTRUSIVE IGNEOUS ROCKS

CLASSIFICATION OF IGNEOUS ROCKS

		Acid	Intermediate	Basic
		Decreasing silica content →		
Decreasing particle size (Faster cooling rcte) ↓	Intrusive	Granite	Diorite	Gabbro
	Extrusive	Rhyolite	Andesite	Basalt

volcanic debris	Ash, Cinders, Blocks, Bombs Composition varies widely

METAMORPHIC ROCKS

Rocks formed under one set of temperature, pressure, and chemical conditions and then exposed to a different set of these conditions may undergo structural and chemical changes, without melting, that produce rocks with different textures and new minerals. This process is known as metamorphism (change in form). Metamorphic rocks are formed deep beneath the earth's surface by the great stresses and high pressures and temperatures associated with mountain building. They are also formed by the intrusion of magma into rock, particularly at the place of contact where the temperatures are high. The study of metamorphic rocks provides valuable information about temperatures and pressures at great depths. Laboratory studies of the stabilities of minerals at temperatures and pressure similar to those within the earth are essential.

METAMORPHIC MINERALS form only at the high temperatures and pressures associated with metamorphism. Among these are kyanite, staurolite, sillimanite, andalusite, and some garnets. Other minerals—the olivines, pyroxenes, amphiboles, micas, feldspars, and quartz—may be found in metamorphic rocks, but are not necessarily the result of metamorphism. These minerals, formed during crystallization of igneous rocks, are stable at high temperatures and pressures and may remain unchanged during metamorphism of the rock. All minerals, however, are stable only within certain limits of pressure and temperature. Thus the presence of some minerals in rocks indicates the approximate temperatures and pressures at which the rocks were formed.

RECRYSTALLIZATION is the change in particle size of minerals during metamorphism. Small gray calcite crystals in limestone, for example, change to large white crystals in marble. Both temperature and pressure contribute to recrystallization. High temperatures allow the atoms and ions in solid crystals to migrate, thus reorganizing the crystals. High pressures cause solution of crystals at their contacts and deposition in the pore spaces between them.

FOLIATION is a layering in metamorphic rock. It occurs when a strong compressive force is applied from one direction to a recrystalling rock. This causes the platy or long crystals of such minerals as mica and tourmaline to grow with their long axes perpendicular to the direction of the force. The result is a banded, or foliated, rock, the bands showing the colors of the minerals that form them. Rocks subjected to uniform pressure from all sides or lacking minerals with distinctive growth habits will not be foliated. Slate is a very fine-grained foliate. Phyllite is a coarse foliate, schist coarser, and gneiss very coarse. Marble is commonly a nonfoliate.

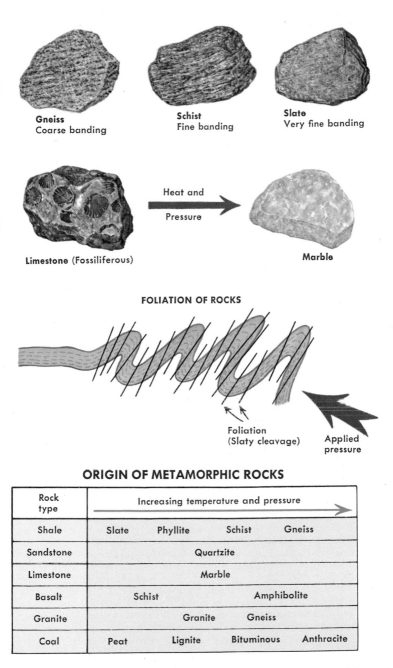

Gneiss
Coarse banding

Schist
Fine banding

Slate
Very fine banding

Limestone (Fossiliferous)

Heat and
Pressure

Marble

FOLIATION OF ROCKS

Foliation
(Slaty cleavage)

Applied
pressure

ORIGIN OF METAMORPHIC ROCKS

Rock type	Increasing temperature and pressure			
Shale	Slate	Phyllite	Schist	Gneiss
Sandstone	Quartzite			
Limestone	Marble			
Basalt	Schist		Amphibolite	
Granite		Granite	Gneiss	
Coal	Peat	Lignite	Bituminous	Anthracite

11

SOLID-STATE REACTION is one of the important mechanisms of metamorphism. It is a chemical reaction between two minerals without either of them melting. In the process atoms are exchanged between the minerals, and new minerals are formed. Consider the minerals quartz and calcite. Each is stable alone at high temperatures. Together in a siliceous limestone, they do not change at low temperatures, but at high temperatures they react with one another and form the metamorphic mineral wollastonite. The chemical equation of the reaction is: SiO_2 (quartz, solid) $+ CaCO_3$ (calcite, solid) $\succ CaSiO_3$ (wollastonite, solid) $+ CO_2$ (carbon dioxide, gas). Many complex high-temperature reactions take place among minerals, and each mineral assemblage produced is a clue to the temperature and pressure at the time of metamorphism.

METASOMATISM is a drastic change in the bulk chemical composition of a rock that often occurs during metamorphism. It is due to the introduction of chemicals from other rocks. Water can transport these chemicals rapidly over great distances. Because of the role played by water, metamorphic rocks generally contain many elements that were absent from the original rock and lack some that were originally present. The introduction of new chemicals is not necessary for recrystallization and solid-state reaction to take place, but it does speed up metamorphic processes.

CONTACT METAMORPHISM describes the chemical changes that take place when magma is injected into cold rock (country rock). These changes in the rock are greatest wherever the magma comes in contact with it, for temperatures are highest at this boundary and decrease with distance from it. Around the igneous rock formed by the cooling of the magma is a metamorphosed zone called a contact metamorphic aureole (halo). Aureoles are important in the study of metamorphism because a single rock type may show all degrees of metamorphism from the contact area to the unmetamorphosed country rock some distance away. Formation of important ore minerals may occur by metasomatism at or near the contact; limestone is particularly susceptible to this type of mineralization.

REGIONAL METAMORPHISM, in contrast to contact metamorphism, involves changes in great masses of rock over wide areas. The high temperatures and pressures in the depths of the earth are the cause. If the resulting metamorphosed rocks are uplifted and exposed by erosion, they may cover many thousands of square miles. Their mineralogy and texture provide important information about mountain building and earth processes. The metamorphism, however, destroys features that would have revealed the rock's previous history. Recrystallization destroys fossils and sedimentary textures; solid-state reaction and metasomatism change the original compositions.

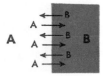

On heating, A and B diffuse through the solids across crystal boundaries.

A new crystal, the compound AB, forms between the crystals.

Diffusion continues until the original crystals are fully reacted.

THE SOLID-SOLID REACTION A + B ⟶ AB

Augen Gneiss

Potassium feldspar crystals formed by introduction of potassium

Quartz veinlet in Mica Schist formed by introduction of silica

CONTACT METAMORPHIC ZONE

Limestone

Shale

Intrusive Granite

Contact metasomatic ore deposits

CONTACT METAMORPHISM NEAR SMALL INTRUSIVE

Intensely folded sedimentary layers metamorphed to Schists and Gneisses

Granite Core

REGIONAL METAMORPHISM ACROSS MANY MILES

SEDIMENTARY ROCKS

All rocks disintegrate slowly as a result of mechanical and chemical weathering. Rock particles—in the form of clay, silt, sand, and gravel—and dissolved materials are transported by the agents of erosion (water, ice, and wind) to new locations, generally at lower elevations, and deposited in layers. The deposited particles eventually become cemented together, forming **clastic** sedimentary rocks. The dissolved materials may precipitate as crystals that accumulate in layers in oceans and lakes and are cemented to form **chemical** sedimentary rocks.

Sedimentary rocks provide abundant information about the most recent half-billion years of the earth's history. They contain in fossil form the preserved remains of evidences of ancient animals and plants. The manner in which particles of sediment are worn and deposited, the relationships of the different layers, the color and composition, the presence of ripple marks or raindrop impressions—these are among the features that enable geologists to reconstruct ancient landscapes and climates as well as the general sequence of geologic events.

MECHANICAL WEATHERING is the breakdown of rock into particles without changing the identities of the minerals in the rock. Ice is the most important agent of mechanical weathering. Water percolates into cracks and pore spaces, freezes, and expands. The force exerted is sufficient to widen cracks and break off pieces, in time disintegrating the rock. Heating and cooling of the rock, with resulting expansion and contraction, also helps. Mechanical weathering contributes further to the breakdown of rock by increasing the surface area exposed to chemical agents. The breakdown of rocks and erosion of the fragments has been greatly accelerated over the past several centuries by the activities of man through farming and construction.

CHEMICAL WEATHERING is the wearing down of rock by chemical reaction. In this process the rock's minerals are changed into finely divided products that can easily be carried away. Air and water are involved in the many complex chemical reactions, which include oxidation, hydrolysis, hydration, and solution. Igneous minerals are unstable under normal atmospheric conditions, those formed at higher temperatures being more readily attacked than those formed at lower temperatures. Igneous minerals are commonly attacked also by water, particularly acid or alkaline solutions. All the common rock-forming igneous minerals (except quartz, which is very resistant) are changed in this way to clay minerals and chemicals in solution. Silica is leached from silicate minerals and removed as a colloidal material that can be deposited later as opal or chert. Clay, quartz, colloidal silica, and chemicals in solution—the common products of weathered rocks—are the building materials of the sedimentary rocks.

SIZES OF SEDIMENTARY ROCK PARTICLES

Name of particles	Diameter of particles 25.4mm = 1 inch	Sedimentary Rock
Boulders Cobbles Pebbles Granules	Greater than 256 mm 64-256 mm 4-64 mm 2-4 mm	Conglomerates (rounded) and Breccias (angular)
Very coarse sand Coarse sand Medium sand Fine sand Very fine sand	1-2 mm ½-1 mm ¼-½ mm ⎫ Sandy ⅛-¼ mm 1/16-1/8 mm ⎭	Sandstones
Coarse silt Fine silt	1/64-1/16 mm ⎫ Gritty 1/256-1/64 mm ⎭	Siltstones and Mudstones
Clay	Less than 1/256 mm Smooth	Shales, Claystones

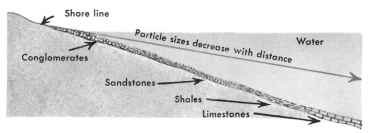

Shore line

Particle sizes decrease with distance

Water

Conglomerates

Sandstones

Shales

Limestones

SORTING BY WAVE ACTION NEAR SHORELINE

Wave and current action is most vigorous near shore,
finer particles are carried to deeper water

CONSTITUENTS OF SEDIMENTARY ROCKS

Major Constituents		Accessory Minerals	
	Abundant	Less abundant	Less than 1%
Mechanically deposited	Quartz Clay minerals Micas Calcite	K-feldspar Plagioclases Rock fragments	Magnetite Tourmaline Garnet Amphibole Hematite Limonite Others
Chemically deposited	Calcite Dolomite	Quartz (chert) Gypsum, Anhydrate Halite Hematite	

TRANSPORTATION AND DEPOSITION of weathered particles is provided by water, wind, and ice. These agents reduce the size of the particles and deposit them in new localities. Sediments dropped by streams form alluvial fans, flood plains, deltas, and deposits in lakes and oceans (p. 15). Winds may move large amounts of sand and smaller particles. Glaciers transport and deposit great quantities of rock materials. Composition of the sediments provides clues to the nature of the original, or source, rock. Differences between successive layers indicate changes that have occurred with time.

PRECIPITATED SEDIMENTS are made up of sodium, potassium, calcium, magnesium, chloride, fluoride, sulfate, carbonate, and phosphate ions. Because these chemicals are very soluble in water, they can be removed from existing rock in solution. Once dissolved in water, they may be precipitated by inorganic processes in oceans or lakes, or may be extracted by living organisms. Calcite ($CaCO_3$), for example, will precipitate from solution in warm waters and settle to the bottom, where it consolidates into limestone rock. Corals, mollusks, and algae also remove $CaCO_3$ from solution. Halite and other very soluble salts normally precipitate only from bodies of water that have no outlet after evaporation produces a saturated solution. Limestone is by far the most abundant precipitate, but salt, gypsum, and phosphate deposits are common.

CLASSIFICATION of sedimentary rocks begins with the broad divisions of clastic and chemical rocks (p. 14), though there is no clear distinction between the two processes of deposition. Chemical rocks are classified on the basis of composition as salt, gypsum, limestone, chert, phosphate rock, nitrate beds, borate beds, etc. Specialized characteristics may be noted by such modifying terms as fossiliferous (containing fossils) and nodular (lumpy).

Classification of clastic rocks is more complex because of many variables. Particle sizes (average and range of sizes), composition of the particles, the cement, the matrix (smaller particles in the spaces among larger grains)—all must be considered. The Wentworth scale is a broad classification based on average particle size. Shale or mudstone, siltstone, sandstone, and conglomerate are names given to rocks with particle sizes ranging from very fine to very coarse. Shales, which consist mainly of clay materials with very fine grains of quartz and feldspar, are generally classified further only on the basis of composition and bedding. Coarser clastics are classified according to composition and particle sizes. Orthoquartzite is a very pure quartz sandstone; arkose, a sandstone with quartz and abundant feldspar; graywacke, a sandstone with quartz, clay, feldspar, and metamorphic rock fragments. The classification on the facing page provides only general terminology, without the specialized names or the descriptive adjectives that are commonly used.

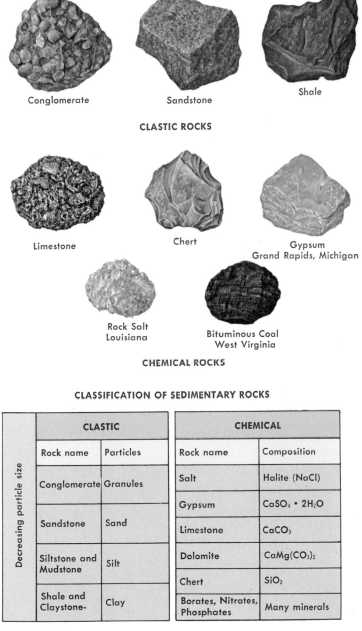

Conglomerate

Sandstone

Shale

CLASTIC ROCKS

Limestone

Chert

Gypsum
Grand Rapids, Michigan

Rock Salt
Louisiana

Bituminous Coal
West Virginia

CHEMICAL ROCKS

CLASSIFICATION OF SEDIMENTARY ROCKS

	CLASTIC		CHEMICAL	
	Rock name	Particles	Rock name	Composition
Decreasing particle size	Conglomerate	Granules	Salt	Halite (NaCl)
			Gypsum	$CaSO_4 \cdot 2H_2O$
	Sandstone	Sand	Limestone	$CaCO_3$
	Siltstone and Mudstone	Silt	Dolomite	$CaMg(CO_3)_2$
			Chert	SiO_2
	Shale and Claystone-	Clay	Borates, Nitrates, Phosphates	Many minerals

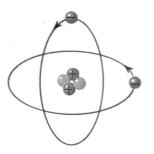

Proton

Neutron

Electron

HELIUM ATOM consists of a nucleus of 2 protons and 2 neutrons with 2 electrons revolving around it.

ATOMIC STRUCTURE OF MINERALS

Like all matter, minerals are composed of the minute particles called atoms. The kinds of atoms in minerals and the patterns in which these atoms are arranged determine the composition and characteristics of minerals. It is obviously important therefore to understand at least generally the atomic structure of minerals.

ATOMS are the building blocks of matter. Too small to be seen even under the most powerful electron microscopes, they are made up of three kinds of even smaller particles—the proton, neutron, and electron. Protons and neutrons are located at the center of the atom, forming its nucleus. Protons carry a positive electrical charge of fixed magnitude $(+1)$; neutrons carry no charge. Electrons revolve around the nucleus and carry a fixed negative charge (-1) equal in magnitude to the charge of the proton. A single, isolated atom is electrically neutral: its number of electrons (negative charges) is equal to its number of protons (positive charges).

ELEMENTS are fundamental substances such as iron, carbon, and oxygen. There are over 100 different elements, each having atoms of a particular kind. The identity of an element is determined by the number of protons in the nucleus of its atom, this number being known as the atomic number. Carbon, for example, has 6 protons in its atom. The number of neutrons is not fixed; the carbon atom may have from 5 to 10 neutrons, but whatever the number the atom remains chemically the same. An element's atomic weight is the sum of the number of protons and neutrons in its atom. Atoms of the same element with different atomic weights are called Isotopes. A carbon atom with 6 neutrons, the most abundant isotope of carbon, is indicated by $_6C^{12}$ (6 being the atomic number, 12 the atomic weight). Carbon with 8 neutrons is $_6C^{14}$. Some isotopes are not stable; carbon 14 ($_6C^{14}$), for example, is radioactive, disintegrating at a fixed rate into an isotope of nitrogen ($_7N^{14}$).

ELECTRONIC CONFIGURATION describes the distribution of electrons in an atom and explains why the atoms of some elements react with others, and why some do not react at all. The electrons occupy space near the nucleus in well-defined energy levels (represented by concentric shells). Electrons in the level nearest the nucleus have the lowest energy, and those farthest from the nucleus have the highest level. Each level can hold a limited number of electrons. The innermost level may contain 2 electrons, the second shell 8, the third shell 18, and each succeeding shell may contain as many as 32.

The number of levels occupied by electrons depends on the number of electrons in the atoms of a particular element. As the number of electrons increases, levels are filled from the innermost one outward. No outer level, however, contains more than 8 electrons; when it contains that number, it remains unfilled until the next level is started, making it the outer one. Elements are very stable if the outer energy levels of their atoms have 8 electrons (a stable octet). Elements react easily if the outer shells of their atoms are not filled or do not have 8 electrons. The atoms of these elements tend to react with other atoms and lose, gain, or share the number of electrons necessary to fill the outer shell or obtain the stable octet.

ELECTRONIC CONFIGURATION OF THIRTEEN ELEMENTS

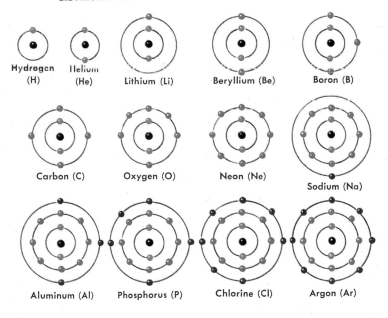

Hydrogen (H) Helium (He) Lithium (Li) Beryllium (Be) Boron (B)

Carbon (C) Oxygen (O) Neon (Ne) Sodium (Na)

Aluminum (Al) Phosphorus (P) Chlorine (Cl) Argon (Ar)

PERIODIC TABLE arranges the elements in a way that shows relationships among them. Long before the electronic structures of atoms became known, the Russian chemist Dmitri Mendeleev and others observed that some elements exhibit remarkably similar chemical behavior. Lithium (Li), sodium (Na), potassium (K), rubidium (Rb), and cesium (Cs), for example, are very metallic, and their atoms tend to lose their electrical neutrality, becoming ions with one positive charge each. These elements, the alkali metals, were therefore grouped together. Fluorine (F), chlorine (Cl), bromine (Br), and iodine (I) are very nonmetallic, and their atoms form ions with one negative charge each. These elements, the halogens, were also grouped together. From such observations the periodic classification of the elements was devised. Gaps in the original periodic table have been filled with elements discovered since that time. Indeed, the development of the periodic table aided materially in the search for unknown elements.

Elements in vertical columns of the periodic table have similar characteristics. Elements listed horizontally have characteristics that grade from left to right. The chemical similarity among elements in vertical columns (groups) and the gradation among elements in horizontal columns (periods) are due to the electronic configurations of the outer levels of their atoms (p. 19). A detailed summary of the chemical characteristics of the groups and periods cannot be given here, but the following generalizations can be noted:

• Elements on the left side (the metals) can attain a stable octet (8 electrons in the outer level) by giving up electrons and becoming positive ions. Those on the right side (nonmetals) can attain a stable octet by gaining electrons and becoming negative ions. Those in the center (semimetals) can attain a stable octet by either giving up or gaining electrons.

• If a metal atom and a nonmetal atom are brought together, the metal atom gives up one or more electrons which the nonmetal atom accepts. The resulting oppositely charged ions are held together by a strong electrostatic force (the ionic bond, p. 22).

• If metal atoms are brought together, neither holds the outer electron (or electrons) strongly, so a metallic crystal is formed, with free electrons moving among the atoms (the metallic bond, p. 26).

• If nonmetal atoms are brought together, both atoms attract extra electrons strongly; they "share" their outer electrons, forming free neutral molecules (the covalent bond, p. 24).

• If semimetals are brought together, their atoms form crystals that have properties between metals and nonmetals (intermediate bonds, p. 28).

• Transition metals (elements 21-30, 39-48, 57-80, and 89-103) are chemically similar to other metals in the same period. Their outer levels have about the same electronic configuration, but their inner levels differ.

PERIODIC TABLE OF THE ELEMENTS

Chemical similarity is pronounced in the groups (vertical columns). Chemical properties vary systematically in the periods (horizontal rows).

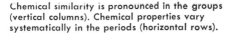

1	2				Transition Elements							3	4	5	6	7	8
1 H	2																2 He
3 Li	4 Be											5 B	6 C	7 N	8 O	9 F	10 Ne
11 Na	12 Mg											13 Al	14 Si	15 P	16 S	17 Cl	18 Ar
19 K	20 Ca	21 Sc	22 Ti	23 V	24 Cr	25 Mn	26 Fe	27 Co	28 Ni	29 Cu	30 Zn	31 Ga	32 Ge	33 As	34 Se	35 Br	36 Kr
37 Rb	38 Sr	39 Y	40 Zr	41 Nb	42 Mo	43 Tc	44 Ru	45 Rh	46 Pd	47 Ag	48 Cd	49 In	50 Sn	51 Sb	52 Te	53 I	54 Xe
55 Cs	56 Ba	57 La	72 Hf	73 Ta	74 W	75 Re	76 Os	77 Ir	78 Pt	79 Au	80 Hg	81 Tl	82 Pb	83 Bi	84 Po	85 At	86 Rn
87 Fr	88 Ra	89 Ac															

Elements 57-71 are the lanthanides, commonly called the rare earths; elements from 89 on are the actinides. All the lanthanides are chemically similar, as are the actinides.

ELECTRONIC STRUCTURES shown by means of concentric shells with the electrons in "orbits" around the nucleus are informative about the energy levels of the electrons but may convey a misleading impression of the structure of an atom. One of the difficulties in defining the structure of an atom is that any experimental method used to determine the position or momentum of an electron will change one or both. The fact that the momentum and position of an electron cannot be determined simultaneously is stated by the Heisenberg uncertainty principle and has forced physicists to use statistical methods to define the structure of the atom.

ELECTRONIC ORBITALS define the locations of the electrons in statistical terms. An electron in the s-subshell of any electronic shell is most likely in a spherical volume of space about the nucleus at any given time. The spherical s-orbital defines the space in which it is probable that the electron is located. Only two electrons, with opposite spins, may be in the s-orbital of a given electron shell. An electron in the next higher subshell is most likely in the volume of space defined by a "dumbbell" shaped p-orbital. Electrons in the d-subshell occupy the "double dumbbell" orbital volume; those in the f-subshell occupy even more complex orbitals.

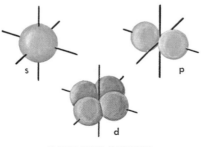

ELECTRONIC ORBITALS

THE IONIC BOND is the force that holds together an atom of a distinctly metallic element to an atom of a distinctly nonmetallic element. Atoms of the more metallic elements have an outer level with 1, 2, or 3 electrons and the next shell with 8. The outer electrons, in general, are not held tightly and can be removed by the energy available in ordinary chemical reactions. The sodium (Na) atom, for example, has 11 electrons—2 in the inner shell, 8 in the second shell, and 1 in the outer shell. The atom achieves stability by losing the outer electron; the second shell with its 8 electrons then becomes the outer shell. Since 1 negative charge is thereby removed, the atom has 1 excess positive charge in the nucleus and is called a positive ion, or cation. It is assigned the symbol Na^{+1}.

Atoms of distinctly nonmetallic elements, in contrast, have 5, 6, or 7 electrons in the outer shell. These atoms obtain the stable octet of outer electrons by attracting electrons from other atoms. The fluorine (F) atom has 9 electrons—2 in the inner shell and 7 in the outer shell; when it borrows 1 electron, it becomes negatively charged. The atom is then called a negative ion, or an anion, and is assigned the symbol F^{-1}.

If an atom of sodium (a metal) and an atom of fluorine (a nonmetal) are brought together, the fluorine atom removes the loosely held electron from the outer shell of the sodium atom. The two atoms thus become oppositely charged. The singly charged cation Na^{+1} and anion F^{-1} are attracted to each other (unlike charges attract one another). The attractive force is the ionic bond.

Not all ions are singly charged. The magnesium (Mg) atom must lose 2 electrons and the oxygen (O) atom must gain 2 electrons to achieve a stable octet; the cation Mg^{+2} and the anion O^{-2} are thus doubly charged. The resulting bond between the 2 ions is therefore about twice as strong as between 2 singly charged ions. In a similar fashion, the aluminum atom forms the Al^{+3} ion and the nitrogen atom forms the N^{-3} ion.

In ionic bonds, the attractive forces are exerted in all directions so that every cation becomes completely surrounded by anions and vice versa. The result is that three-dimensional periodic arrays—crystals—are formed. Ions are packed closely in crystal structures and the bonds are non-directional. Only two requirements must be fulfilled: (1) cations are surrounded by as many anions as is geometrically possible, and (2) the total number of positive charges in the crystal equals the total number of negative charges.

Ions with different charges can form crystals. For example, 2 Fe^{+3} ions combine with 3 O^{-2} ions, forming Fe_2O_3 (hematite), an electrically neutral crystal. Some elements can exist as two or more ions; ferrous iron, Fe^{+2}, and ferric iron, Fe^{+3}, are both common in minerals. Chemically the 2 ions behave as different elements. In magnetite, Fe_3O_4, the formula is written as $Fe^{+2}Fe_2^{+3} O^{-2}$. The Fe^{+2} ion is surrounded by 6 O^{-2} ions, and the Fe^{+3} ion is surrounded by 4 O^{-2} ions.

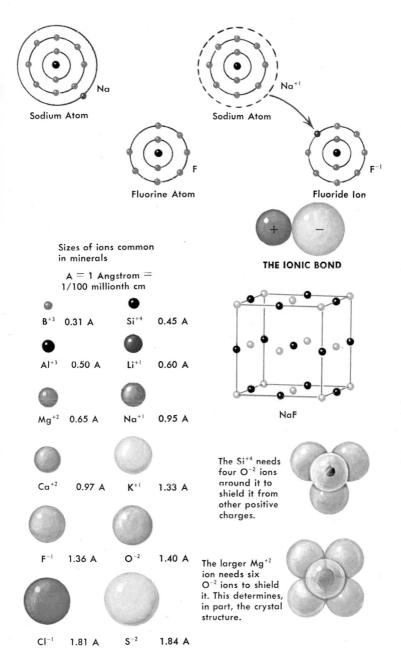

Sodium Atom

Sodium Atom Na^{+1}

Fluorine Atom F

Fluoride Ion F^{-1}

THE IONIC BOND

Sizes of ions common in minerals

A = 1 Angstrom = 1/100 millionth cm

B^{+3} 0.31 A

Si^{+4} 0.45 A

Al^{+3} 0.50 A

Li^{+1} 0.60 A

Mg^{+2} 0.65 A

Na^{+1} 0.95 A

Ca^{+2} 0.97 A

K^{+1} 1.33 A

F^{-1} 1.36 A

O^{-2} 1.40 A

Cl^{-1} 1.81 A

S^{-2} 1.84 A

NaF

The Si^{+4} needs four O^{-2} ions around it to shield it from other positive charges.

The larger Mg^{+2} ion needs six O^{-2} ions to shield it. This determines, in part, the crystal structure.

THE COVALENT BOND is the force that holds together two non-metallic atoms. When brought together, these tend not to become ionized because both have a strong affinity for electrons. A chlorine atom (Cl), for example, has 17 electrons—2 in the inner shell, 8 in the second shell, and 7 in the outer shell. It will attain a stable octet of electrons in the outer shell if a metallic atom is available to donate an electron. Because the energy required to remove 7 electrons is very large, the atom cannot attain a stable octet easily by giving up electrons. Two chlorine atoms brought together can attain stable octets, however, if each shares one of its electrons with the other. The resulting Cl_2 molecule, having reached stability, has little tendency to form a crystal at ordinary temperatures. The sharing of a pair of electrons is the covalent bond. The bonding energy is gained by reduction of the energies of the two atoms. This bond forms the strongest of all chemical ties.

Atoms that form covalent bonds with other atoms will participate in the formation of crystals of five types, depending on the number of covalent bonds the atoms can form.

1. Molecular crystals consist of distinct covalently bonded molecules held together by van der Waals forces (very weak forces due to small residual charges on the surface of molecules). Solidified gases, such as carbon dioxide (dry ice) and chlorine, and many organic compounds form crystals of this type.

2. Ionic crystals are formed by complex ions that are held together by covalent bonds and carry charges. The ammonium ion, $(NH_4)^{+1}$, and the carbonate ion, $(CO_3)^{-2}$, within which the bonds are covalent, are bonded to Cl^{-1} and Ca^{+2} ions by ionic forces in NH_4Cl (sal ammoniac) and $CaCO_3$ (calcite).

3. Crystals with chain structures, as in the semimetals selenium and tellurium, have atoms linked together with covalent bonds in infinite chains. The chains are held together by van der Waals and metallic bonding forces.

4. Crystals with sheet structures have atoms linked together with covalent bonds in infinite layers. The layers are held together by van der Waals forces. Graphite is the most striking example of this type structure.

5. Crystals with framework structures form if an atom can share electrons with four adjacent atoms, as in diamond. Bonding is infinite in all three dimensions, and the crystal can be considered a molecule of infinite size.

Since covalent bonds are very strong, their distribution determines the characteristics of a crystal. It explains the splitting of graphite into flaky layers, the hardness of diamond, the low melting point of dry ice, and the formation of complex ions in solution—$(HCO_3)^{-1}$ and $(NH_4)^{+1}$, for example.

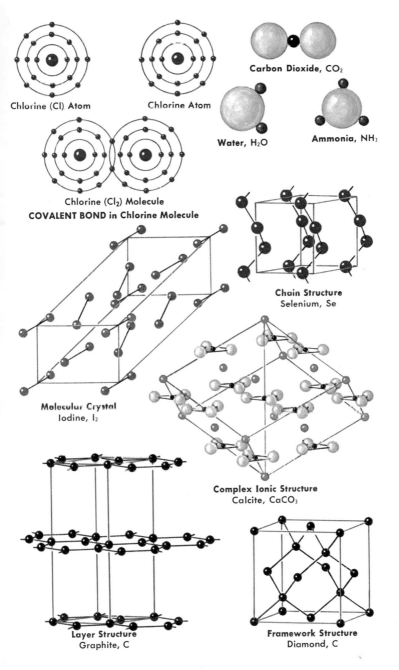

Chlorine (Cl) Atom

Chlorine Atom

Carbon Dioxide, CO₂

Water, H₂O

Ammonia, NH₃

Chlorine (Cl₂) Molecule

COVALENT BOND in Chlorine Molecule

Chain Structure
Selenium, Se

Molecular Crystal
Iodine, I₂

Complex Ionic Structure
Calcite, CaCO₃

Layer Structure
Graphite, C

Framework Structure
Diamond, C

THE METALLIC BOND is the force that binds together atoms of the metallic elements. These include the first, second, and third groups of the periodic table and the transition elements (p. 20). All have one feature in common: 1, 2, or 3 electrons in the outer energy level of their atoms. If two of these atoms with loosely held outer electrons are brought together, covalent bonds (p. 24) or ionic bonds (p. 22) normally do not form. Rather, a crystal forms in which the atoms are packed together tightly. The outer electrons are attached to no atom in particular, but are free to move through the structure. The bonding energy may be thought of as an attraction between the positive nuclei of the atoms and the negative charges of the free-moving electrons. The metallic bond imparts certain very distinct characteristics to the crystal.

The atoms in metal crystals are packed together to occupy the maximum amount of space, resulting in three common metal structures: the face-centered cubic and the close-packed hexagonal, representing the densest packing possible; and the body-centered cubic, only slightly less dense. More complex structures are the result of some covalent or ionic character of the bond. Because the free electrons can move through the structure, an electric potential applied across the crystal results in high electrical conductivity. Heat applied to one side of the crystal will cause agitation of the electrons, which impart energy to other electrons and conduct heat through the crystal very rapidly. Shearing forces applied to metal crystals can cause gliding along atomic planes without brittle fracture; the metals are thus ductile and malleable. Metallic luster and opacity are caused by interaction of light with the free electrons.

Metallic bonds can be important in nonmetallic crystals, notably the sulfides. Sulfides can be described best by considering their structure as covalently bonded. Many sulfides, however, contain metal atoms that are not well-shielded from one another, and some metallic bonding has developed between them. A striking result of this is the strong metallic luster of some sulfides as in galena (PbS), which looks much like lead metal (Pb). Because the covalent bond generally controls the mechanical properties of sulfides, they are brittle.

There are few crystals in the crust of the earth that consist of metals held together by metallic bonds. Because metals easily lose their outer electrons and because there are abundant non-metals available to accept them—oxygen (O) and sulfur (S), for example—most metals are chemically combined in compounds. Thus metal oxides and sulfides are the common minerals. The native metallic elements (pp. 70-75) are those which do not form very stable oxides or sulfides, such as gold (Au), silver (Ag), and platinum (Pt), or are formed in oxygen-deficient environments, such as copper (Cu) and iron (Fe). The core of the earth, however, most probably consists of nearly pure metal, iron (Fe) and nickel (Ni), the inner portion of which is liquid. Many meteorites have a composition similar to the earth's core.

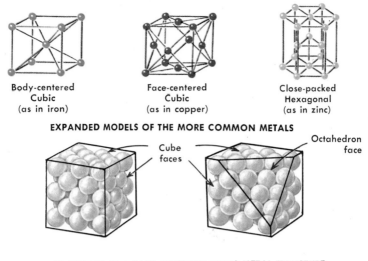

Body-centered
Cubic
(as in iron)

Face-centered
Cubic
(as in copper)

Close-packed
Hexagonal
(as in zinc)

EXPANDED MODELS OF THE MORE COMMON METALS

Cube faces

Octahedron face

Cube faces

PACKING MODELS OF A FACE-CENTERED CUBIC METAL STRUCTURE

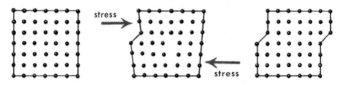

stress →

← stress

Sheer stress causes deformation of metal crystals; metallic bonds can be broken and reformed without fracturing the crystal, as shown. This characteristic is known as malleability.

← HEAT

metal bar

Heat flows rapidly through metals because free electrons can transmit energy when they collide. Electricity is transmitted by electrons also and metals are good electrical conductors.

Galena, shows metallic luster, but is brittle.

Lead, with metallic bonding, is malleable.

INTERMEDIATE BONDS lie between the three extremes represented by metallic, covalent, and ionic bonds: (1) very metallic elements (with low electron affinity) combine to form metallic bonds; (2) very nonmetallic elements (with high electron affinity) combine to form covalent bonds; and (3) very metallic and very nonmetallic elements combine to form ionic bonds. If all elements were either very metallic or very nonmetallic, all crystals would be characterized by one of these three bond types. But elements range between the very metallic elements of Group I of the periodic table and the very nonmetallic elements of Group VII.

Elements near the middle of the periodic table are neither metals nor nonmetals. In silicon (Si), for example, the energy required to remove or to add the 4 electrons needed for a stable octet is quite large. Pure silicon therefore forms a covalent crystal essentially identical with that of diamond. But silicon has a distinct metallic luster and is opaque, indicating that the shared electrons are less strongly bonded than in diamond. In other words, the bond is covalent, but with some metallic character. When silicon is in chemical combination with oxygen (O), a covalent bond is possible, with the 4 outer electrons of silicon pairing with electrons of oxygen. An ionic bond, with Si^{+4} bonded to O^{-2} ions, is another possibility. Silicon is neither strongly metallic nor nonmetallic, so the bond is neither purely covalent nor purely ionic. The Si-O bond appears to involve both ionization and electron sharing; it is stronger than the pure ionic bond, but not as strong as a pure covalent bond. Moreover, Si and O can form complex ions of finite or infinite size (pp. 154-7).

In general, elements that appear near each other on the left side of the periodic table (the metals) form metallic crystals (crystals with metallic bonds). Elements near each other on the right side of the table (the nonmetals) form molecules with covalent bonds. Elements near each other in the central portion form covalent crystals with some metallic characteristics, or metal crystals with some covalent characteristics.

Elements on opposite sides form ionic crystals. The farther apart they are the more ionic character the bond will have. The sodium-oxygen (Na-O) bond, for example, is quite ionic. The magnesium-oxygen (Mg-O) bond is ionic, but shows more covalent character. The aluminum-oxygen (Al-O) bond is even less ionic, and the silicon-oxygen (Si-O) bond appears to be about half ionic and half covalent.

Crystals having chemical bonding intermediate between covalent and metallic are particularly important in solid state electronics. Silicon, for example, is a semi-conductor; it conducts electricity only above a certain voltage, when doped with small impurities, and then in only one direction. This rectifying characteristic makes silicon useful in transistors. Understanding this concept of intermediate bonds is particularly important because the physical and chemical behavior of crystals is directly related to the bond type.

CHARACTERISTICS OF BONDS AND CRYSTALS

Compounds	Bond Characteristics	Crystal Characteristics
Li, Na, K, Rb, Cs, Ca, Sr, Ba	nearly 100% metallic	metallic, ductile, very conductive
Be, Mg, Al, Ga, In, Tl, Ge, Sn, Pb, Sb, Bi, Po	metallic with some covalent characteristics	metallic, somewhat brittle, conductive
B, C, Si, P, As, Se, Te, I, At	covalent with some metallic characteristics	semimetallic, brittle, poorly conductive
N, O, S, F, Cl, Br	covalent	molecules, low melting point, very poorly conductive
KCl, NaF, etc.	ionic	ionic, poorly conductive, dissolves to ions in H_2O
Al_2O_3, SiO_2, SnO_2	ionic-covalent	Intermediate characteristics
As_2S_3, Sb_2S_3, etc.	covalent-metallic	Intermediate characteristics

METALLIC CHARACTERISTIC OF THE ELEMENTS

Elements are increasingly metallic from top to bottom and from right to left in the periodic table.

Least Metallic

Transition Elements

Most Metallic

Elements 57-71 are the lanthanides, commonly called the rare earths; elements from 89 on are the actinides. All the lanthanides are chemically similar, as are the actinides.

THE HYDROGEN BOND is a special bond formed only between the hydrogen atom and other atoms. It is a weak polar bond. In minerals the hydrogen bond is virtually limited to compounds containing the water molecule H_2O (hydrated compounds) or the hydroxyl ion $(OH)^{-1}$ (hydrous compounds). The effects of water and hydroxyl on the mineral structure are pronounced. Their absence or presence in minerals is due to varying conditions of formation and provides valuable clues to geologic environments.

THE HYDROGEN ATOM consists of a single proton in the nucleus and a single electron. The oxygen atom has 8 protons in the nucleus and 8 electrons—2 in the inner shell and 6 in the outer shell. Thus 2 additional electrons are needed for a stable outer octet. A mixture of hydrogen and oxygen at high temperature will react explosively and produce water, H_2O, in which an electron pair is formed between 1 oxygen and 2 hydrogen atoms. This provides both hydrogens with a stable pair of electrons. Complex measurements indicate, however, that this covalent bond is modified by some repulsion between the hydrogen atoms. This means that to some extent the hydrogens must be positively charged, or ionized.

An important feature of the water molecule is that, because of the small size of the hydrogen atom, the hydrogen atoms are "buried" in the oxygen's cloud of electrons. The water molecule is essentially spherical, therefore, and nearly the same size as the oxygen atom.

THE HYDROXYL ION consists of a hydrogen atom and an oxygen atom with a shared electron. To complete its stable octet, the oxygen accepts another electron from a metallic atom, with which it forms an ionic bond. As in the water molecule, the hydrogen atom is "buried" in the oxygen ion, and the hydroxyl ion is nearly the same size as the oxygen ion (1.40 A).

Although both the hydroxyl ion bonded to a metal ion and the water molecule are neutral, they are polar. The site of hydrogen carries a small positive charge (because of its proton), and the other side of the water molecule or hydroxyl ion carries a small negative charge. Both, therefore, behave as small magnets. The positive pole can exert an attractive force on the electrons of other atoms. This is the hydrogen bond. It is the force that holds water molecules together as ice and that contributes to bonding in any hydrated or hydrous mineral. In minerals, the hydrogen bond between hydroxyl ions and adjacent oxygen ions is particularly important.

Hydrogen bonding between complex organic molecules is responsible for many chemical processes that take place in living organisms, controlling metabolic activity, reproduction and hereditary traits, etc. Biochemical processes are enormously more complex, however, than the chemical processes involved in mineral crystallization.

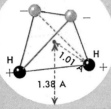

The tetrahedral distribution of charge on water molecule, H_2O

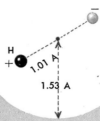

The linear distribution of charge on the hydroxyl ion $(OH)^-$

Both water and the hydroxyl ion act as magnets, the positive poles being attracted to negative poles in crystal structures.

Tetrahedral arrangement of hydrogen bonds about the water molecule

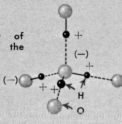

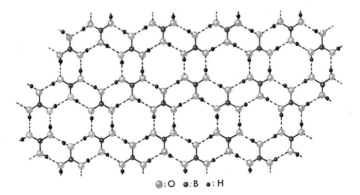

⬤:O ⚫:B ●:H

A layer of the boric acid structure showing hydrogen bonding between $B(OH)_3$ molecules

31

SOLID SOLUTIONS

Minerals are commonly thought of in terms of a simple chemical formula that is fixed and unchangeable. Forsterite, for example, is Mg_2SiO_4; anhydrite, $CaSO_4$. These formulas, however, rarely express the exact chemical compositions; they are only approximations. Compositions of most minerals are not fixed, but vary between definite limits. This variation is referred to as **solid solution** or **diadochy.**

In a crystal, an element occupies a particular position because of its characteristics, notably the size of its ion or the type of chemical bonding. Many elements have atoms with electronic configurations similar to those of other elements, however, and are chemically similar; either element may be stable in the same structural position. In silicates, for example, the position occupied by the silicon ion may also be occupied by the aluminum ion, which is nearly the same size and has the same electronic configuration. But it is important to remember that the proportion of silicon and aluminum in these positions is not fixed, but depends on how much of each element was present during the mineral's crystallization.

Similarly, magnesium (Mg) and iron (Fe) may be present in a number of minerals in identical positions, and the Mg:Fe ratio may have any value. Regardless of the Mg:Fe ratio, however, the crystal structures of minerals with variable amounts of Mg and Fe are essentially identical. The elements Mg and Fe are said to be "substituting" for one another in the structure. Minerals with identical structures, but with differing chemical compositions are said to be **isomorphous.**

COMPLETE SOLID SOLUTION occurs when two elements can substitute for one another completely in a mineral. Olivines are excellent examples of this type of solid solution. Forsterite, Mg_2SiO_4, and fayalite, Fe_2SiO_4, have identical structures because the ions Mg^{+2} and Fe^{+2} are very nearly the same size and are chemically similar. If both iron and magnesium are present in the environment of crystallization, a single mineral with a composition intermediate between forsterite and fayalite will form. This mineral is considered to be a "solution" of fayalite in forsterite. Because minerals with all possible intermediate compositions can be formed, the formula is written $(Mg,Fe)_2SiO_4$, and the general name, regardless of Mg:Fe ratio, is olivine.

The olivines are referred to as members of a complete solid-solution series between the Mg end member, forsterite, and the Fe end member, fayalite. Mg-rich olivines are light green, and Fe-rich olivines are dark green or black. Olivines with intermediate compositions are intermediate in color. Their other physical characteristics also vary between those of the end members. In such minerals, either pure end member is seldom found, though they may be synthesized in the laboratory.

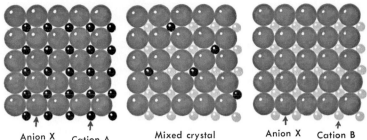

Anion X Cation A
Pure AX

Mixed crystal
(A,B)X

Anion X Cation B
Pure BX

Solid solution may be complete if substituting ions are nearly the same size and are chemically similar.

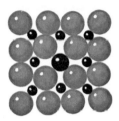

If substituting ions are much larger or smaller, the structure is distorted and weakened. Only limited substitution is possible.

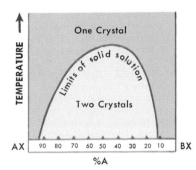

The limits of solid solution increase with increasing temperature

Ordered

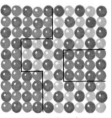

Disordered

Solid solutions of certain compositions, as 50-50, tend to order. These ordered crystals have properties like compounds rather than intermediates. Disordered solid solutions have properties intermediate between those of the end members.

LIMITED SOLID SOLUTION occurs in a mineral when two elements that substitute for one another do not form all possible proportions. Solid solution in the olivines is complete because iron and magnesium ions are nearly the same size. But in many other minerals, only limited amounts of one ion can substitute for another because the two ions are of different size. Anhydrite, $CaSO_4$, and barite, $BaSO_4$, are chemically similar, but have different structures because the Ba^{+2} ion is considerably larger than the Ca^{+2} ion. Chemical analyses, nevertheless, show that anhydrite can contain up to 8 percent $BaSO_4$, and barite can contain up to 6 percent $CaSO_4$. Substitution of more ions cannot take place without disrupting the structure at either end of the series; the amount of solid solution is thus limited to narrow ranges of composition at each end. Similarly, potassium feldspar ($KAlSi_3O_8$) and sodium feldspar ($NaAlSi_3O_8$) do not form complete solid solutions except at high temperatures. To sum up, the extent of solid solution is governed by the relative sizes and chemical similarity of the substituting elements and by the temperature.

COUPLED SUBSTITUTION involves simultaneous substitutions that balance one another. It results in a more complex type of solid solution, either complete or limited, that occurs in some minerals. The feldspars are an example. Calcium feldspar (anorthite, $CaAl_2Si_2O_8$) and sodium feldspar (albite, $NaAlSi_3O_8$) have the same basic structure. The calcium ions (Ca^{+2}) in calcium feldspar are in the same positions as the sodium ions (Na^{+1}) in sodium feldspar; the aluminum ions (Al^{+3}) and silicon ions (Si^{+4}) occupy identical positions in both minerals, but occur in different proportions. Because the Ca^{+2} ion and the Na^{+1} ion are nearly the same size and can substitute for each other, a complete solid-solution series is possible between the two minerals. But because Ca^{+2} and Na^{+1} have different charges, the substitution of one for the other would unbalance the neutrality of the over-all structure and could therefore not occur. It becomes possible, however, because Al^{+3} and Si^{+4} can also substitute for one another. When both substitutions occur simultaneously, neutrality is maintained. The "exchange" can be written $Ca^{+2} + Al^{+3} : Na^{+1} + Si^{+4}$, 5 charges replacing 5 charges. Feldspars with all possible Ca:Na ratios can be found, with the Al:Si ratio varying accordingly. Members of this complete solid-solution series, all similar in appearance, are called plagioclase feldspars.

IMPORTANCE OF SOLID SOLUTION is great because most minerals exhibit it at least to a limited extent. Variations in composition provide valuable clues to temperatures and chemical environments during formation. Solid solution explains why some elements are nearly always found in trace amounts in minerals and are thereby "camouflaged." Analyses of trace elements in rocks can lead to the discovery of concentrations of valuable elements in other nearby rocks.

34

SOLID SOLUTION IMPORTANT IN MINERALS

OLIVINES—Substitution of Mg^{+2}, Fe^{+2}, and Mn^{+2} (pp 164-165)
  Forsterite—Fayalite: $Mg_2SiO_4 - Fe_2SiO_4$ Complete
  Forsterite—Tephroite: $Mg_2SiO_4 - Mn_2SiO_4$ Complete

CORDIERITES—Substitution of Fe^{+2} for Mg^{+2} (pp 172-173)
  Cordierite—Iron Cordierite: $Mg_2Al_3(Si_5Al)O_{18} - Fe_2Al_3(Si_5Al)O_{18}$
  Complete

MELILITES—Coupled substitution of $Si^{+4} + Mg^{+2}$ for $2Al^{+3}$
    (pp 168-169)
  Gehlenite—Akermanite: $Ca_2Al(SiAl)O_7 - Ca_2MgSi_2O_7$ Complete

PYROXENES—Substitution of Mg^{+2} and Fe^{+2} (pp 174-181)
  Diopside-Hedenbergite: $CaMg(SiO_3)_2 - CaFe(SiO_3)_2$ Complete
  Ca: (Mg, Fe) varies over limited range
  Intermediate compositions are called pigeonite
  Augite is intermediate but has Al^{+3} in composition

AMPHIBOLES—Substitution of Mg^{+2} and Fe^{+2} Complete (pp. 182-185)
  Tremolite-Ferroactinolite: $Ca_2Mg_5(Si_8O_{12})(OH)_2 - Ca_2Fe_5(Si_8O_{12})$
  $(OH)_2$ Complete
  Intermediate compositions are called ferrotremolite,
  Tremolite-actinolite, or simply actinolite

MICAS—Coupled substitution of $3Fe^{+2}$ for $2Al^{+3} + $ Vacancy
    (pp 194-197)
  Muscovite-Annite: $KAl_2(Si_3Al)O_{10}(OH)_2 - KFe_3(Si_3Al)O_{10}(OH)_2$
  Biotite is $K(Mg, Fe)_3(Si_3Al)O_{10}(OH)_2$ with Mg:Fe less than 2:1
  Phlogpite is $K(MgFe)_3(Si_3Al)O_{10}(OH)_2$ with Mg:Fe greater than
    2:1

PLAGIOCLASE FELDSPARS—Coupled substitution of $Ca^{+2} + Al^{+3}$ for
    $Na^{+1} + Si^{+4}$ (pp 218, 219)
  Albite-Anorthite: $NaAlSi_3O_8 - CaAl_2Si_2O_8$ Complete intermediate
  compositions are called plagioclase

ALKALI FELDSPARS—Substitution of K^{+1} and Na^{+1} (pp 214-217)
  Microcline (Sanidine)—Albite: $KAlSi_3O_8 - NaAlSi_3O_8$
    Complete above about $670°C$
    At ordinary temperatures about 20% substitution of Na^{+1} in
    microcline and about 5% substitution of K^{+1} in Albite

CRYSTALS

Crystals have always been of great interest because of their intrinsic beauty. To mineralogists they also provide clues to the arrangement of atoms within a mineral and offer an important means of mineral identification. Only a few minerals, such as opal and silica glass, lack a crystal structure.

A CRYSTAL is a solid material with a regular internal arrangement of atoms. Because of this orderly internal arrangement, a crystal may form the smooth external surfaces called faces. Faces represent planes in the crystal on which there are closely spaced atoms and on which growth can occur rapidly. The orientation of these faces is an excellent indicator of the internal symmetry possessed by the structure and of the nature of the chemical bonding (p. 20).

Though we usually think of crystals as having faces, these smooth surfaces are not always present. The majority of mineral specimens, in fact, consist of aggregates of crystals that have not developed faces because of mutual interference during formation (anhedral crystals). Crystals exhibit well-developed faces (euhedral crystals) only when they grow unhampered, as on the walls of an open cavity, on a surface, in the early stages of crystallization of an igneous melt, or in other unusual circumstances.

THE LAWS OF SYMMETRY prove that all crystals can be placed in 1 of 7 categories called systems. These 7 systems can be further subdivided into 32 symmetry classes and 230 groups based on internal arrangement. Treatment of these subdivisions constitutes the study of crystallography, an intriguing subject that is beyond the scope of this book. On the following pages the 7 crystal systems, with illustrations of representative crystal types, are treated briefly.

THE FACES observed on a crystal can be described in terms of regular polyhedra—for example, cubes and octahedra. All faces of a polyhedron are of equal size and like shape and are referred to collectively as a crystallographic form. Some crystals exhibit only one form; others exhibit two or more forms, some having large faces and others small.

In a real crystal, the different faces of a form are not necessarily the same size or shape. Growth of the crystal in one direction more rapidly than in other directions results in malformed polyhedra, often producing forms that are difficult to recognize. But the angles between the faces of a form are the same, regardless of the distortion. Real crystal faces, moreover, may not be perfectly smooth and flat, but may have undulations or other irregularities, caused by irregular growth or by interference from external forces during crystal growth. Illustrations of real crystals (pp. 70-257), compared with ideal crystals (pp. 38-43), demonstrate many of these distortions and irregularities.

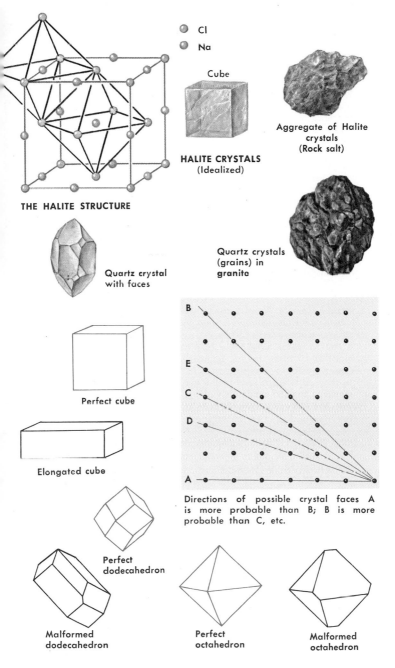

Cl
Na

Cube

HALITE CRYSTALS
(Idealized)

Aggregate of Halite
crystals
(Rock salt)

THE HALITE STRUCTURE

Quartz crystal
with faces

Quartz crystals
(grains) in
granite

Perfect cube

Elongated cube

Perfect
dodecahedron

Malformed
dodecahedron

Perfect
octahedron

Malformed
octahedron

Directions of possible crystal faces A
is more probable than B; B is more
probable than C, etc.

CRYSTAL SYSTEMS

The seven systems in which crystals can be placed are defined by three or four imaginary axes of equal or unequal length that intersect at the center of any perfect, undistorted crystal form. The lengths of the axes and the angles between them define the crystal's shape. Because crystals can grow to different sizes, the relative lengths of the axes, as indicated by a specimen, are commonly noted. Absolute lengths are determined by X-ray analysis and are expressed in terms of a "unit cell," the smallest unit exhibiting all the symmetry of the whole crystal. Axes of unit cells are a few hundred-millionths of a centimeter long (2.54 cm = 1 in.). Unit cells of many minerals, with atoms shown, are illustrated on pages 70-257 of this book.

CUBIC SYSTEM includes crystals with three mutually perpendicular axes of equal length. The axes are conventionally drawn perpendicular to the three pairs of faces of a cube or through the three pairs of corners of an octahedron. Common forms of cubic crystals include the tetrahedron (4 faces), the cube (6 faces), the octahedron (8 faces), the dodecahedron and pyritohedron (both 12 faces), the cubic trapezohedron (24 faces), and the hexoctahedron (48 faces). A cubic crystal may have any one of these forms or any combination of these forms. Some of the possible combinations are shown on the opposite page. To make recognition of the forms more readily apparent to the reader, different colors have been used for the different crystallographic forms. It is important to note the different appearances caused by different degrees of development of two forms.

Crystals containing several forms may appear to be quite complex. Important in recognizing which forms are present are the angles between the individual faces of a form, since they are always the same regardless of the number of forms present or the degree of malformation. Many minerals commonly develop some forms and not others, which makes the recognition of forms important in mineral identification. The characteristic forms and the kinds of malformations are referred to as the mineral's habit. (Discussion of habits on p. 46.)

Many chemical compounds, including minerals, form cubic crystals. Simple compounds (those with one, two, or three elements) tend to crystallize in either the cubic system or the hexagonal system (p. 40). Included in this category are some of the elements, simple sulfides, oxides, and halides. More complex compounds—the silicates, for example—tend to form crystals with axes of unequal lengths, which places them in the other crystal systems. Garnets and some feldspathoids, among others, are notable exceptions; in spite of their relative complexity, they form cubic crystals.

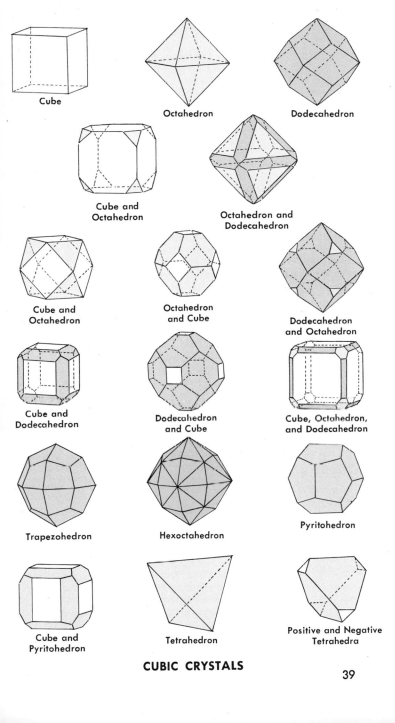

Cube

Octahedron

Dodecahedron

Cube and
Octahedron

Octahedron and
Dodecahedron

Cube and
Octahedron

Octahedron
and Cube

Dodecahedron
and Octahedron

Cube and
Dodecahedron

Dodecahedron
and Cube

Cube, Octahedron,
and Dodecahedron

Trapezohedron

Hexoctahedron

Pyritohedron

Cube and
Pyritohedron

Tetrahedron

Positive and Negative
Tetrahedra

CUBIC CRYSTALS

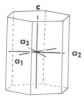

HEXAGONAL SYSTEM includes crystals with 4 axes, 3 of which are of equal length and lie in a plane with angles of 120° between them. The fourth axis, the unique or hexagonal axis (c), is perpendicular to the plane of the other 3 and may be of any length.

A common form is the hexagonal prism, consisting of 6 faces parallel with the unique axis. The ends of the crystal may be two parallel faces (pinacoid or base), two pyramids (dipyramid), or a combination of a flat face and a pyramid. There are many possible pyramid and dipyramid forms, differing only in the angle that the faces of each make with respect to the unique axis. Dihexagonal prisms and dipyramids are common; these actually are double forms with 12 faces rather than 6. Hexagonal crystals are common among minerals consisting of simple compounds, moderately common among more complex minerals. As with the cubic forms (p. 39), the different hexagonal forms are shown in different colors on the facing page. Some hexagonal crystals show only three-fold symmetry (trigonal) but are not rhombohedral (below).

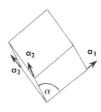

RHOMBOHEDRAL SYSTEM, sometimes included as a subdivision of the hexagonal system, includes crystals with 3 axes of equal length, as in the cubic system, but with an angle (α), other than 90°, between them. The most common forms are the rhombohedron, the trapezohedron, and the scalenohedron. Hexagonal prisms may be found in combination with these forms, illustrating the close connection between the two systems. In the crystals of the rhombohedral system, however, a view along the unique axis (c) shows only 3 equal faces (3-fold symmetry) rather than 6 or 12 faces (6-fold symmetry) as in crystals of the hexagonal system. Calcite and quartz are among the minerals that crystallize in the rhombohedral system. The different forms are shown in different colors on the facing page. Though hexagonal forms, such as the prism and the dipyramid, may be present, as on quartz (facing page), the overall symmetry is rhombohedral.

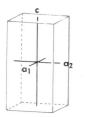

TETRAGONAL SYSTEM includes crystals with 3 mutually perpendicular axes, 2 of which are of equal length. The forms include prisms, pyramids, and dipyramids, which are distinguished from similar forms of the hexagonal system by having 4 faces rather than 6. Other forms are the tetragonal trapezohedron, the sphenoid, and the tetragonal scalenohedron. Ditetragonal forms, with 8 faces, are also common. Tetragonal crystals are well represented among the common rock-forming minerals; the scapolites, rutile, and zircon are some more common examples. The different forms are shown in different colors.

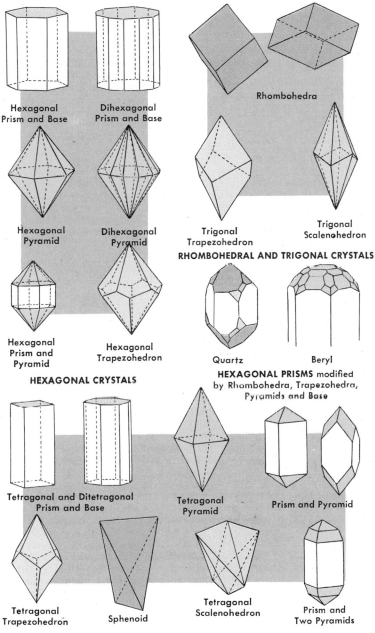

Hexagonal
Prism and Base

Dihexagonal
Prism and Base

Rhombohedra

Hexagonal
Pyramid

Dihexagonal
Pyramid

Trigonal
Trapezohedron

Trigonal
Scalenohedron

RHOMBOHEDRAL AND TRIGONAL CRYSTALS

Hexagonal
Prism and
Pyramid

Hexagonal
Trapezohedron

Quartz

Beryl

HEXAGONAL CRYSTALS

HEXAGONAL PRISMS modified
by Rhombohedra, Trapezohedra,
Pyramids and Base

Tetragonal and Ditetragonal
Prism and Base

Tetragonal
Pyramid

Prism and Pyramid

Tetragonal
Trapezohedron

Sphenoid

Tetragonal
Scalenohedron

Prism and
Two Pyramids

TETRAGONAL CRYSTALS

41

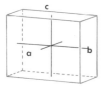

ORTHORHOMBIC SYSTEM includes crystals with 3 mutually perpendicular axes of different lengths. The most common forms are the pinacoid (2 parallel faces), the dipyramid, the prism, and the dome, illustrated on the facing page. There are no axes about which the crystal can be rotated to show 3-fold, 4-fold, or 6-fold symmetry. Two-fold axes are present, however, and the orthorhombic crystals therefore are commonly blocky in appearance.

If one dimension of the unit cell is notably longer than the other two, the crystal as a whole may be needlelike; if one unit-cell dimension is notably shorter than the other two, the crystal may be platy. The crystal shape is determined also by the rate of growth in different directions, hence is not necessarily related to the relative unit-cell dimensions. Compounds that are chemically complex or have complex bonding commonly crystallize in the orthorhombic system.

MONOCLINIC SYSTEM includes crystals with 3 axes of different lengths, 2 of which are perpendicular. The third is not perpendicular to the plane of the other 2, but makes an angle β with one so the crystals characteristically look like orthorhombic crystals that have been deformed in one direction. The forms are the same as the orthorhombic forms, but the shape of the unit cell is different. Common forms are prisms oriented in the direction of one of the axes and pinacoids parallel with the planes of any two axes. As in the orthorhombic system, compounds that are chemically complex or have complex bonding are commonly monoclinic, including many of the micas, pyroxenes, and sulfides. Gypsum, some of the borates, and many less common minerals are also monoclinic.

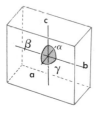

TRICLINIC SYSTEM includes crystals with 3 axes of different lengths, none of which is perpendicular to the others. The angles between the axes are called α, β, and γ. This low symmetry restricts the possible crystal forms to (1) the pinacoid, consisting of two parallel planes, and (2) the pedion, consisting of one plane with no parallel planes. The pinacoid and the pedion may have any orientation with respect to the chosen axes. The low symmetry makes the structures of triclinic crystals difficult to determine and difficult to describe in terms of crystal forms. As in the monoclinic system, complex compounds commonly crystallize in the triclinic system. Among the few triclinic minerals are the most common rock-forming mineral groups—the feldspars and some of the micas, the clays, and kyanite. Many minerals can exist in either monoclinic or triclinic forms depending on the composition or temperature formation.

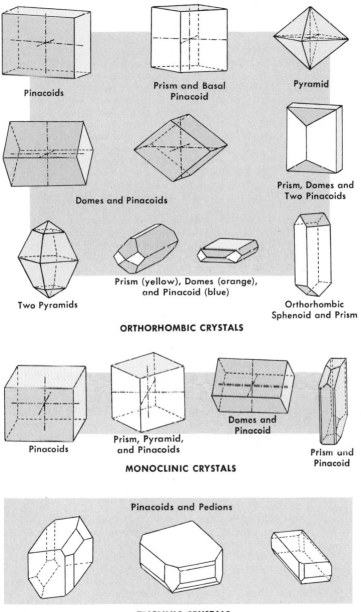

Pinacoids

Prism and Basal
Pinacoid

Pyramid

Domes and Pinacoids

Prism, Domes and
Two Pinacoids

Two Pyramids

Prism (yellow), Domes (orange),
and Pinacoid (blue)

Orthorhombic
Sphenoid and Prism

ORTHORHOMBIC CRYSTALS

Pinacoids

Prism, Pyramid,
and Pinacoids

Domes and
Pinacoid

Prism and
Pinacoid

MONOCLINIC CRYSTALS

Pinacoids and Pedions

TRICLINIC CRYSTALS

TWINNED CRYSTALS

A twinned crystal is a single crystal that appears to consist of two or more crystals grown together along a plane. This composition plane between "twins" is not to be confused with the boundary between two separate crystals, where mismatching is severe and the crystals are easily separated. Twinning occurs most commonly when the growth of a crystal undergoes slight shifts in direction. These changes in direction of growth are the result of spacings among the atoms on certain atomic planes within the crystal. Twinning is common and easily observed among crystals with well-developed faces and, less obviously, within individual grains in rocks.

CONTACT TWINS are twinned crystals that exhibit growth in two directions from a common plane. The composition plane, generally obvious, is always a direction of a possible crystal face. In addition to the simple contact twins just described are cyclic contact twins. These show repeated twinning at regular intervals, the crystal growth occurring along a "circular" path, as in chrysoberyl. Though there are many possible twin planes for any given crystal, twinning normally occurs along a few planes on which favorable interatomic spacings are found. Twinning is said to occur in accordance with a number of "twin laws," which are described in terms of the crystallographic orientation of the composition planes.

PENETRATION TWINS appear to have one twin penetrating the other. They form for the same reasons as any other twin, but are more conveniently described in terms of twinning about an axis instead of a plane. The total number of probable twin axes for a given compound is normally small. Fluorite commonly is found as interpenetrating cubes; pyrite forms "iron crosses" of twinned pyritohedra. Staurolite and arsenopyrite are very commonly found as twinned crosses, consisting of penetrating prisms that are popular collectors' items. Orthoclase is commonly twinned in several ways, one of which is called the Carlsbad twin, shown on the opposite page. Quartz prisms also form penetration twins. In many cases one of the twin members is much smaller than the other.

POLYSYNTHETIC TWINS are contact twins that are repeated on the same composition plane at close intervals across the entire crystal. A number of minerals, notably the plagioclase feldspars, exhibit this type of twinning. The thickness of individual twin members is roughly the same in a given crystal, but may vary from one crystal to another. Polysynthetic twinning on a very small scale (even microscopic) commonly results in a striated appearance, as in most labradorite feldspars. Scattering of light by closely spaced polysynthetic twin planes causes a strange glow called chatoyance.

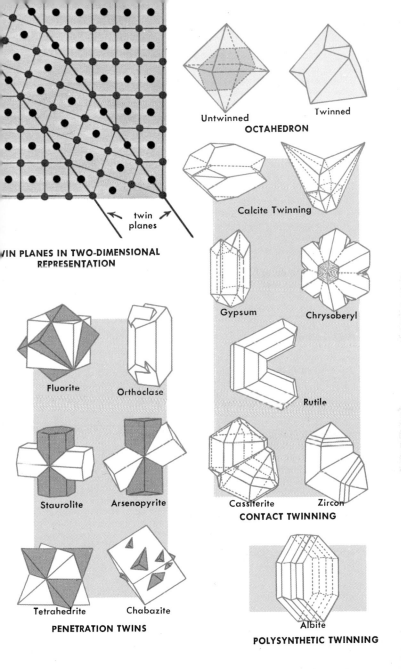

TWIN PLANES IN TWO-DIMENSIONAL REPRESENTATION

twin planes

Untwinned   Twinned
OCTAHEDRON

Calcite Twinning

Gypsum   Chrysoberyl

Rutile

Cassiterite   Zircon
CONTACT TWINNING

Fluorite   Orthoclase

Staurolite   Arsenopyrite

Tetrahedrite   Chabazite
PENETRATION TWINS

Albite
POLYSYNTHETIC TWINNING

45

CRYSTAL HABIT

The shape and size of a crystal or aggregate of crystals are called the "habit." In a sense, the faces of a perfectly formed crystal constitute its habit, but characteristic malformations of crystals, particle-size distributions in aggregates, and many other distinctive features are included in the term. Many factors, some not completely understood, influence crystal habit. Internal structure, hence the rate of crystal growth in different directions, is most important. Also important are certain factors present during crystal formation: space available, temperature, chemical environment, and kind of solution. The names that are given to the differing habits of crystals are usually readily understandable descriptive terms. In learning to identify minerals, it is important to become familiar with the variations in all of the properties observed in a given mineral. Among the most important of these is habit, which may be diagnostic.

PRISMATIC HABIT refers to a crystal that is appreciably longer in one direction than in the other two. Some of the most distinctive crystals —those of stibnite, asbestos, satinspar (a fibrous gypsum), pyroxenes, and amphiboles—are prismatic. The prismatic habit is described more specifically by such names, generally self-explanatory, as: columnar, bladed, acicular (needlelike), and fibrous. Minerals that have prismatic crystals are commonly found in aggregates of several minerals.

GRANULAR HABIT most commonly describes aggregates composed of mineral grains cemented together, regardless of the habit of the crystals. Descriptive names include coarse granular, fine granular, and powdery granular.

LAMELLAR HABIT refers to a crystal that is appreciably shorter in one direction than in the other two. There are various types. Tabular crystals are thick. Micaceous crystals are very thin, and the lamellae can be peeled off along cleavage planes. Foliated crystals have thin lamellae that may be folded and distorted, as in talc. Plumose crystals are featherlike. Lamellar aggregates are groups of lamellar crystals cemented together.

HABITS OF AGGREGATES of minerals may be quite different from the habits of the constituent minerals. Aggregate habits are described by a number of readily understood terms, such as globular, massive, and massive earthy. Special names are given to those with special structures.

Aggregates composed of groups of minerals, generally prismatic and radiating from a common center to form a spherical surface, may be given one of several names, depending on the size of the groups. If the groups are small (less than $\frac{1}{8}$ in. each), the structure is called oölitic; if they are larger ($\frac{1}{8}$-$\frac{1}{2}$ in.), it is pisolitic; if they are even larger, it is botryoidal, or mammalary. If the groups are large and kidney-shaped, the structure is said to be reniform. Stellate aggregates are radiating groups resembling stars. Dendritic aggregates consist of individual crystals of any habit that grow in a branching manner. Geodes are groups of crystals, banded or radiating, that grow inside a cavity in a rock. Subsequently, these groups may be dissolved out as rounded masses that, when split, reveal the interior crystals. Drusy crystals are aggregates of small, well-formed crystals that grow on the surface of a rock or along fractures in rocks.

COLUMNAR
Rhodonite

BLADED
Actinolite

FIBROUS
Serpentine

PRISMATIC HABITS

ACICULAR
Rutile

TABULAR
Zoisite

MICACEOUS
Muscovite

LAMELLAR HABITS

RADIATING LAMELLAE
Pyrophyllite

MASSIVE EARTHY
Hematite

BOTRYOIDAL
Hematite kidney ore

MINERAL AGGREGATES

DENDRITIC
Silver

DRUSY
Chabazite

47

IMPERFECTIONS IN CRYSTALS

To understand crystals, it is necessary to consider ideal specimens, with all faces of a form present and with each face perfectly shaped and planar. It is customary also to consider the internal structure as perfect, with no irregularities in the arrangement of the constituent atoms. In reality, however, there is no such thing as a perfect crystal. Crystal faces may be misshapen, etched, curved, and irregular in a number of ways. These deviations are of great value in mineral identification and provide clues to the mode of formation. Irregularities in the atomic arrangement are responsible, too, for many of the observed mechanical and electrical properties of crystals.

SURFACE IRREGULARITIES are imperfections that can be observed on the surface of a crystal, either with the unaided eye or with a microscope. They may be formed during crystal growth or at any time subsequently. Striated surfaces, as observed on quartz crystals and many other mineral crystals, result from changes in growth direction during crystallization. Close examination reveals many very small parallel faces on any groove or ridge. Curved crystal faces, as on calcite and dolomite, are of similar origin. Many crystal faces are pitted or etched, commonly a result of chemical activity on a smooth face, but in some cases formed during crystal growth. The faces of some crystals have depressions, which themselves have faces; others have small elevations (vicinal prominences), also with faces. They are the result of unequal growth rate on the crystal face.

STRUCTURAL DEFECTS are internal deviations from the idealized perfect crystal. Every crystal of visible proportions contains many millions of such defects. These "mistakes" of nature are important in determining the behavior of crystals; indeed, they seem to be instrumental in promoting crystal growth. Point defects involve only one structural position, their effects extending no more than a few atoms beyond the defect. Structural vacancies (often inaccurately called lattice vacancies) are sites where an atom is lacking. These defects affect a crystal's optical, electrical, and mechanical properties. Impurity atoms, located at structural positions or in interstitial positions (between the "normal" atoms), also affect the properties. Line defects involve many atoms, essentially in one direction, and are the result of missing rows or planes of atoms. An important defect is the helical, or screw, dislocation. This is formed by continued crystal growth about an edge dislocation. The theoretical strength of a crystal may be many times the measured values in real crystals. The inevitable presence of structural defects provides locations of stress concentration that causes the crystal to yield by fracture or by deformation. The malleability of metals is an important property because dislocations enable the metal to slip, without fracturing.

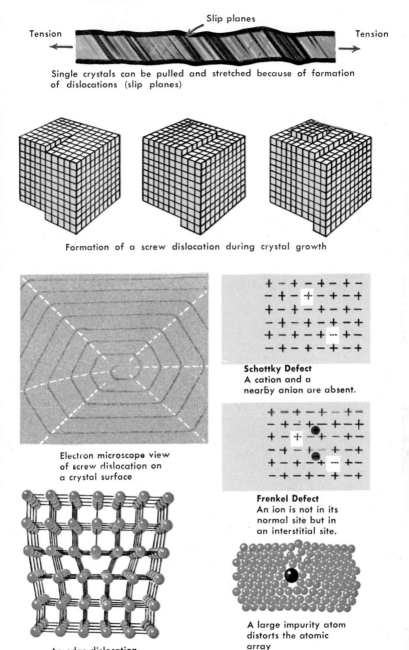

Slip planes

Tension →   ← Tension

Single crystals can be pulled and stretched because of formation of dislocations (slip planes)

Formation of a screw dislocation during crystal growth

Electron microscope view of screw dislocation on a crystal surface

Schottky Defect
A cation and a nearby anion are absent.

Frenkel Defect
An ion is not in its normal site but in an interstitial site.

An edge dislocation

A large impurity atom distorts the atomic array

49

CRYSTAL INCLUSIONS

During or after crystallization, a mineral may be exposed to high temperatures and solutions of other materials for very long periods of time. As a result, some material may be trapped in the crystal or some portions may be dissolved and replaced by a different mineral or even by a nonmineral. Such inclusions commonly occur in minerals. In many cases when one mineral is included in another, it is difficult to determine which mineral formed first. But if this can be determined, the sequence of crystallization and the history of formation of the entire rock can be worked out. Also, certain kinds of inclusions may be characteristic for particular minerals and therefore of value in identification. A few of the many types and varieties of inclusions are discussed below.

RANDOM INCLUSIONS are those inclusions whose shape and orientation are not related regularly to the crystallographic directions, or symmetry, of the crystal. These may simply be bubbles, filled or partially filled with gas or liquid. A study of the fillings can provide valuable information about the chemistry of the environment in which the crystal was formed.

Random inclusions may also be crystalline in structure. Quartz crystals, for example, frequently contain needles of rutile, hematite, asbestos, tourmaline, or other minerals. Aventurine feldspar contains very small particles of hematite, which are responsible for the distinctive reddish color. Sand crystals, which are large calcite crystals found in sandstone, contain high percentages of evenly distributed sand grains (quartz). Nearly any crystal may have inclusions. Some inclusions, such as quartz in staurolite (poikilitic staurolite), are characteristic and aid in identification. Sometimes random inclusions are aggregated within the crystal.

ORIENTED INCLUSIONS occupy spaces along definite crystallographic planes of the host mineral, producing a symmetrical pattern. These intergrowths can form as a result of simultaneous crystallization of the host and a mineral inclusion (eutectic intergrowths). Sometimes, intergrowths are due to the migration of atoms along certain crystallographic planes when some elements separate out of the host mineral at high temperatures. This is the phenomenon of exsolution.

Graphic granite is a good example of a eutectic intergrowth in which microcline contains quartz intergrown along some crystallographic directions. Perthite is a very good example of an intergrowth resulting from exsolution. At high temperatures, albite ($NaAlSi_3O_8$) is completely soluble in microcline ($KAlSi_3O_8$). As cooling occurs, however, albite separates out along the crystallographic planes of microcline, producing a banded appearance. Chiastolite, with inclusions of carbon along certain directions, is a crystal with noncrystalline inclusions.

RANDOM CRYSTALLINE INCLUSIONS

Tourmaline needles in quartz

Rutilated quartz (Rutile needles)

Sand crystals Quartz grains incorporated into calcite during crystal growth

ORIENTED INCLUSIONS

Graphic Granite (Quartz and Potassium Feldspar)

Chiastolite Inclusions in Andalusite

Perthite Na-Feldspar in K-Feldspar

FLUID INCLUSIONS

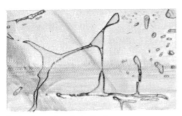

Inclusions in Sphalerite: Rounded areas are gas bubbles. Remainder of inclusions filled with salt water (x14)

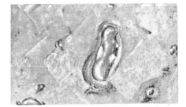

Inclusion of Liquid Mercury in Calcite (x10)

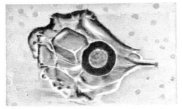

Cubic Salt Crystal and Carbon Dioxide Bubble in Emerald (x195)

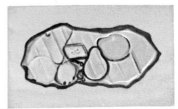

Several Crystals and Gas Bubble (pear-shaped) in Magnesite (x120)

(After photographs by Edwin Roedder, *Scientific American*, Vol. 207, No. 4, p. 38, October, 1962.)

PROPERTIES OF MINERALS

The properties described briefly on the next several pages are important in the identification of minerals and are relatively easy to determine, at least approximately. Many other properties are not discussed because they are uncommon or require elaborate laboratory equipment for their determination.

CLEAVAGE is the tendency of some minerals to break along definite planes when struck with a sharp blow, as with a hammer. The flat cleavage surfaces provide a means of identification and a clue to structure. Cleavage planes parallel possible crystal faces. Breaks occur along these planes rather than through them because in general they consist of closely spaced atoms bound together by forces stronger than those binding atoms in other directions.

Cleavage is described as perfect, good, fair, or poor, depending on the regularity of the break and the luster of the surface. It is also described by the number of cleavage directions and the angle between them, or by the resulting form. Mica, for example, breaks in one direction; a "book" of mica exhibits two parallel cleavage surfaces, but they constitute a single direction. Galena has three directions of cleavage at right angles (cubic cleavage) and thus has six cleavage planes (three pairs in three directions). Other common types of cleavage are shown at right.

Some minerals—halite and calcite, for example—have such well-developed cleavage directions that a break in any other direction is virtually impossible. Other minerals, such as the pyroxenes and amphiboles, have well-developed cleavage in some directions, but break unevenly in other directions.

FRACTURE is the irregular breakage of a mineral. Quartz and beryl, for instance, break unevenly in all directions. There are no observable cleavage planes. Fracture is common in minerals with complex structures in which there are no directions of unusually weak bonding.

Descriptive terms applied to fractures are nearly all self-explanatory. Parting, as exhibited by corundum and garnet, is a poorly developed cleavage in which the parting surfaces are not planes but are irregular. Conchoidal (shell-like) fracturing is common in amorphous materials, such as opal and volcanic glass. Some minerals that crystallize with unusual habits fracture in a manner characteristic of those habits. Satin-spar gypsum and asbestos serpentine, for example, are fibrous in habit and exhibit a fibrous fracture. Uneven, hackly (jagged), splintery, blocky, and prismatic are other terms used to describe fracture. As with all properties, it is desirable to become familiar with names applied to different kinds of fracture and to study examples using real mineral specimens. Fracture characteristics may appear different in different specimens of the same mineral.

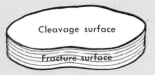

One-directional cleavage
as in micas

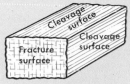

Two-directional cleavage
at right angles as in
Pyroxene

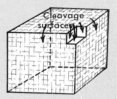

Three-directional cleavage
at right angles (cubic),
as in Halite

Two-directional cleavages
not at right angles, as in
Amphiboles

Three-directional cleavage
not at right angles, as in
Calcite

Octahedral cleavage,
as in Fluorite

Conchoidal fracture
Obsidian

Fibrous fracture
Goethite

Irregular fracture
Quartz

HARDNESS is one of the most characteristic and useful properties of a mineral. It can be defined in many ways, each a measure of the mineral's resistance to a mechanical force. Some of these forces (and the scales that measure them) are: scratching (Mohs), indentation (Knoop), abrasion (Pfaff), and grinding (Rosiwal).

Testing resistance to scratching is a popular method of determining hardness because it requires no special equipment. The Mohs' scale (opposite page) consists of 10 minerals of different hardness. Each is given a number from the softest (talc, 1) to the hardest (diamond, 10). A mineral that can be scratched by quartz (7) but not by microcline (6) has a hardness between 6 and 7. The same mineral will, of course, scratch microcline but will not scratch quartz.

Common objects can also be used as test instruments—the fingernail (2½), a copper coin (3½), a knife blade or common glass (5½), and a metal file (6½). A sharp edge of the test object is moved firmly across a relatively flat surface of the mineral to be tested. Then a moistened fingertip is applied to the surface to remove loose powder and reveal if the surface has been scratched rather than marked.

At best, the scratch test provides a rough estimate of hardness. A weathered or altered mineral may be softened on the surface, and some minerals are harder in one direction than in others.

SPECIFIC GRAVITY is a measure of a mineral's density, or weight per unit volume. Water is assigned a specific gravity of 1 (1 gram per cubic centimeter), and the specific gravities of other materials are compared to it. A mineral fragment that weighs 2.1 times as much as an equal volume of water has a specific gravity of 2.1, assuming there are no pore spaces.

The jolly balance and the beam balance are used for a fairly accurate determination of a mineral's specific gravity. In both, a fragment is weighed first in air and next in water. The difference between the two weights is the weight of an equal volume of water. Dividing this into the weight in air gives the specific gravity, as shown on the facing page.

After practice, a moderately good estimate of specific gravity can be made merely by lifting a mineral fragment. A piece of lead is obviously heavier than a piece of glass of equal size; likewise, the relative weights of other minerals with closer specific gravities can be distinguished. Numerical values, however, cannot be determined in this fashion.

Practice with minerals of known specific gravity is invaluable. Often a rough estimate is sufficient to distinguish between two minerals that are very similar in appearance. Care is needed if the specimen is porous or contains other minerals than the one being investigated; the apparent specific gravity may be misleading. The specific gravity of a mineral varies with composition, as in solid solutions, but in general, the difference can be detected only by careful measurement.

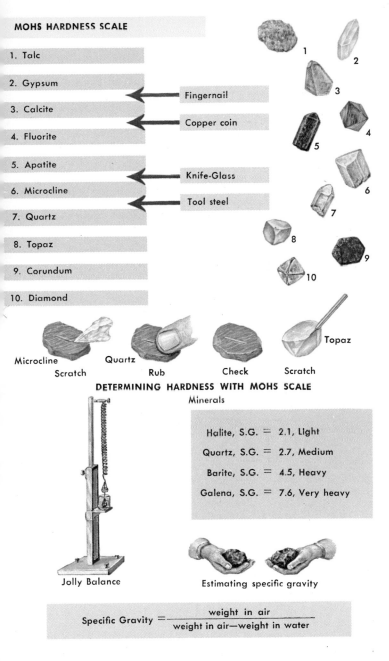

MOHS HARDNESS SCALE

1. Talc
2. Gypsum
3. Calcite
4. Fluorite
5. Apatite
6. Microcline
7. Quartz
8. Topaz
9. Corundum
10. Diamond

Fingernail → (between 2. Gypsum and 3. Calcite)

Copper coin → (between 3. Calcite and 4. Fluorite)

Knife-Glass → (between 5. Apatite and 6. Microcline)

Tool steel → (between 6. Microcline and 7. Quartz)

Topaz

Microcline   Quartz
Scratch   Rub   Check   Scratch

DETERMINING HARDNESS WITH MOHS SCALE

Minerals

Halite, S.G. = 2.1, Light

Quartz, S.G. = 2.7, Medium

Barite, S.G. = 4.5, Heavy

Galena, S.G. = 7.6, Very heavy

Jolly Balance

Estimating specific gravity

$$\text{Specific Gravity} = \frac{\text{weight in air}}{\text{weight in air} - \text{weight in water}}$$

COLOR may be of help in identifying a mineral. Some minerals are invariably the same color (ideochromatic), but most of them occur in a variety of colors (allochromatic). The different colors may be the result of small inclusions of other minerals, of chemical impurities in the structure, or of defects in the structure. Even most allochromatic minerals, however, are likely to be one particular color or one of several colors, and the colors may be useful in interpreting the history of the mineral's formation. The majority of sulfides and sulfosalts are ideochromatic, as are some carbonates. The other classes are generally allochromatic.

Color should always be noted on a freshly broken surface, because staining and corrosion may alter the appearance of an older surface. The kind of staining or corrosion, however, may also be an aid to identification. Sulfides and sulfosalts may be difficult to identify, but the corrosion product, generally a compound of the metals of the sulfide or sulfosalt, is often diagnostic.

LUSTER, like color, is determined by the way light is reflected from a mineral. A particular luster, however, is more likely to be characteristic of a given mineral than is a particular color. Luster is the sheen or gloss of a mineral's surface. The general appearance of the surface, the reflectivity, the scattering of light by microscopic flaws or inclusions, the depth to which light penetrates—these and many other features are responsible for luster. Terms used to describe luster are nearly all common and self-explanatory: earthy (dull), metallic, vitreous (glassy), pearly, resinous, greasy, silky, and adamantine (brilliant, like a diamond).

OPACITY is closely associated with luster. An opaque mineral does not transmit light to any extent. A translucent mineral does transmit light. A transparent mineral not only transmits light, but permits objects to be seen clearly through it. It is worth remembering that all minerals with a metallic luster are opaque.

Opacity may be difficult to determine. A transparent mineral may be translucent if it contains many tiny fractures, numerous inclusions, or has been altered on the surface by weathering. Some minerals may be translucent or transparent only on thin edges or may vary with composition.

STREAK is an important, simple test that can be used to identify some minerals. Minerals with a hardness less than that of an unglazed porcelain plate (5½) will leave a streak of finely powdered mineral when rubbed across the plate. The powder's color is commonly characteristic of the mineral.

Colored minerals generally leave a streak of the same color as the mineral (congruent). Some do not, however. Black hematite, for example, leaves a reddish-brown streak. Weathered minerals may leave a streak that is not characteristic of the mineral but of the alteration product formed on its surface. Other minerals leave either no streak or a white streak that is useless.

56

Azurite and Malachite

Galena

Pyrite

Chalcopyrite

SOME IDEOCHROMATIC MINERALS

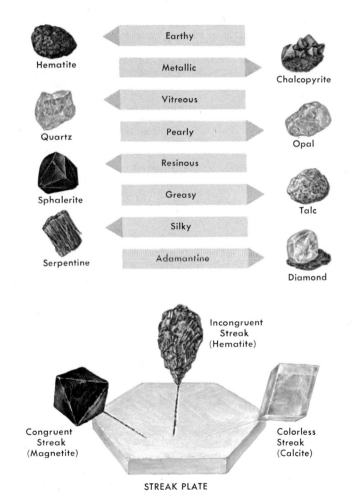

Hematite

Quartz

Sphalerite

Serpentine

Earthy

Metallic

Vitreous

Pearly

Resinous

Greasy

Silky

Adamantine

Chalcopyrite

Opal

Talc

Diamond

Incongruent Streak (Hematite)

Congruent Streak (Magnetite)

Colorless Streak (Calcite)

STREAK PLATE

TASTE may be useful in identifying water-soluble minerals, such as the halides and borates, which have characteristic tastes. Halite, or common salt (NaCl), is familiar to everyone. Sylvite (KCl) is bitterer than halite, but is so alike in taste that it is nearly impossible to distinguish. Fortunately, the water soluble minerals are not toxic, but it is best to avoid tasting minerals except as a final test. Perspiration from the fingers will dissolve enough of the specimen so that the solubility may be detected rather easily without tasting.

SOLUBILITY IN ACIDS is a property that may distinguish one mineral from another. Some minerals, notably the carbonates, are very soluble in acids. Calcite ($CaCO_3$) dissolves rapidly and with vigorous effervescence; dolomite ($CaMg(CO_3)_2$) dissolves more slowly and without effervescence. A small bottle of acid with a dropper can be carried in the field to identify these common carbonates.

Other minerals, such as nepheline, do not dissolve readily in acids, but are altered by them; a characteristic gel forms on their surface. Still other minerals are insoluble.

A mineral's solubility depends on the kind of acid, its strength, and its temperature. Many sulfates are soluble only in concentrated sulfuric acid (H_2SO_4). Other minerals are insoluble in cold hydrochloric acid (HCl), but soluble in hot. Most silicates are soluble only in hydrofluoric acid (HF), an acid so dangerous that it is best to use other tests. Consult an advanced mineralogy text before conducting extensive chemical tests.

MAGNETISM is exhibited by only three common minerals in enough strength to show up in simple tests. Magnetite (Fe_3O_4) is strongly magnetic (may be attracted by a magnet). The magnetite called lodestone is permanently magnetized and so will attract pieces of unmagnetized iron. Pyrrhotite ($Fe_{1-x}S$) and ilmenite ($FeTiO_3$) are less strong magnetic. A magnet is useful in identifying these minerals.

PYROELECTRICITY is the polarization of many crystals when they are heated; that is, negative charge accumulates at one end of the crystal, positive charge at the other end. Tourmaline is one of the more spectacular of these pyroelectric crystals. When heated, it will attract small bits of paper to its ends. Many pyroelectric crystals are only weakly polarized, and this characteristic is then of little use for identification.

PIEZOELECTRICITY is the polarization of some crystals, notably quartz and tourmaline, when subjected to pressure. These crystals also will exert outward pressure if an electric current is applied to them and will vibrate if the current is alternating. The property is of interest because of its importance for electronic and ultrasonic devices, but is of little use for identification.

DETERMINATION OF CRYSTAL STRUCTURES

The structure or arrangement of atoms in a crystal is directly related to the properties of minerals and to their behavior. Prior to 1900, scientists had speculated about the structures of minerals, concluding that all crystals are built of small identical units, called unit cells, arranged in a three-dimensional periodic array. With the discovery of X-rays, a means of determining the arrangements of atoms in unit cells became available.

X-RAYS are electromagnetic waves, like light, with much shorter wavelengths, in the vicinity of an angstrom, 1/100 millionth of a centimeter. Because successive planes of atoms are spaced at distances of one to several angstroms apart, they diffract X-rays in the same way that a ruled grating diffracts light. Using the equation $n\lambda = 2d \sin \Theta$, where n is an integer, 1, 2, 3, n . . . , λ is the wavelength of the incident X-rays, Θ is the angle of incidence, and d is the distance between identical planes, interplanar spacings can be determined by measuring the intensities of X-rays scattered at different angles by the crystal. The crystal scatters X-rays in such a way that the spots or lines are produced on a film or a chart. The size and shape of the unit cell can be determined mathematically from the positions of the spots, and the locations of atoms in the unit cell can be determined from the intensities of the spots or lines only after rather complex and detailed mathematical analysis. It is this coherently scattered X-radiation which is important in crystal structure determination.

The X-ray crystallographer also uses information from other analytical methods. A chemical analysis of a crystal is essential. Valuable clues to the structure can also be obtained with a petrographic miscroscope, from density measurements, from measurement of interfacial angles, from study of pits etched on the crystal by acids, from measurement of pyroelectric and piezoelectric properties, and even from hardness measurements and observation of surface luster and texture.

THE EFFICIENCY with which X-rays are scattered by an atom is directly related to the atomic number. Light elements, such as hydrogen, lithium, etc., are difficult to place within the unit cell, particularly if heavier elements are also present. When the crystal is that of a compound, the problem of determining its structure is even more complex since different atoms do not scatter X-rays equally.

Neutron diffraction, using relatively slow neutrons from reactors, can be used to help determine locations of the lighter elements because the efficiency of the lighter elements in scattering neutrons is in many cases quite high. Electron diffraction, carried out by placing the crystal in the path of the beam of an electron miscroscope, is also very useful, particularly in determining the structure of very small crystals or of their surface films.

LUNAR MINERALS AND ROCKS

The successful manned landings on the moon, beginning with Apollo 11 in 1969, initiated a new era of mineralogic and petrologic research. Rock samples brought from the moon have been studied, and will continue to be, in almost incredible detail. Because the moon lacks an atmosphere, evidence of the initial formation of rocks has not been destroyed by weathering as it has been on the earth. Lunar rocks therefore provide valuable clues to the origin of the moon and the related early stages of formation of the earth.

LUNAR ROCK MATERIALS have been classified on the basis of overall appearance into four types: (A) dark gray, very fine-grained rocks, similar in composition to terrestrial basalts; (B) coarser-grained rocks, with crystals larger than 1 mm. (.04 in.), also similar to basalts; (C) breccias, consisting of fragments of coarse minerals cemented in a matrix of very fine material, similar in appearance to terrestrial breccias, though not formed by stream deposition; and (D) fines, material similar to soil, scooped from the surface.

LUNAR MINERALOGY is much less complex than that of the earth. The most notable difference is the absence of the hydrous minerals, micas and amphiboles, and other minerals that form by precipitation from aqueous solution. The major lunar minerals are calcic plagioclase, $CaAl_2Si_2O_8$, and pyroxene, $(Ca,Mg,Fe)SiO_3$. These, with ilmenite, $FeTiO_3$, make up the bulk of the lunar basalts. Common are olivine, $(Mg,Fe)_2SiO_4$; cristobalite and tridymite, the two high-temperature forms of SiO_2; and pyroxferroite, $CaFe_6(SiO_3)_7$, a new mineral similar to the pyroxenes. Found as accessory minerals (less than 1 per cent) are metallic iron and nickel; troilite, FeS; chromite, $FeCr_2O_4$; ulvospinel, Fe_2TiO_4; two phosphates—apatite, $Ca_5(PO_4)_3(F,Cl)$, and whitlockite, $Ca_9MgH(PO_4)_7$; and a new oxide called armacolite, $(Fe,Mg)Ti_2O_5$. Particularly interesting is a small amount of rock material containing quartz, SiO_2, and potash feldspar $KAlSi_3O_8$, found in one breccia and several samples of fines. These minerals, major components of granite, indicate differentiation of the lunar body, probably in the early stages, in a manner similar to the process by which the crust of the earth developed. Radioactive studies indicate that the lunar rocks crystallized 3.1 to 4.0 billion years ago, corresponding with the ages of the oldest rocks on the earth.

LUNAR GLASS is common. It occurs as a constituent of the basalts, formed by rapid cooling of molten lava at or near the surface, and as small spherical beads mixed with the surface fines, obviously the result of melting and splashing of surface material by the intense force of impacting meteors. Composition varies widely.

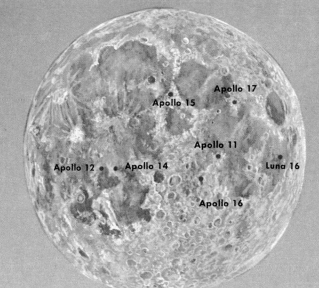

MOON LANDING SITES

LUNAR ROCKS

Type B: Basalt

Type A: Vesicular Basalt

Type C: Lunar Breccia

Type D: Lunar Fines
with Astronaut Footprint

LUNAR GLASSES

61

CHEMICAL COMPOSITION of lunar rocks is, on average, similar to that of terrestrial basalts. Some elements, however, are notably more abundant in the lunar rocks. Among these are titanium and chromium, which are about 10 times as abundant in lunar basalts as in terrestrial basalts. Manganese, zirconium, hafnium, beryllium, and some of the rare earth elements are also more plentiful in lunar rocks. Silicon, aluminum, sodium, and potassium, notably abundant in terrestrial granites, are much less abundant in lunar rocks.

WATER ON THE MOON has been detected in two instances. A crystal of goethite, a hydrated iron oxide, was found by Cambridge University investigators in a breccia from a ridge of the Fra Mauro formation. It has been speculated that the mineral was formed by reaction between iron and water escaping from a volcanic vent, or possibly that some water was present during crystallization of a lunar magma and that the mineral formed directly. Water also has been detected during a moonquake at the same locality by an ion detector left on the surface.

LUNAR HISTORY, though still largely speculative, has been elucidated greatly by observation and photography of the lunar surface and by examination of mineral specimens. It appears quite certain that the moon originated about 4.5 billion years ago, at the same time as the earth, and that the moon, like the earth, was volcanically active. During the first 1.5 billion years or so, the moon was heated, possibly by radioactive disintegration processes, and some differentiation of the body occurred, with lighter materials tending to accumulate in the surface layer. Some evidence suggests that the lunar basins, or maria, are chemically similar to the deep ocean basins of the earth, containing basic, or basaltic, rocks, and that the lunar highlands are somewhat more similar to the continents of the earth, containing some acid, or granitic, rocks. Because the moon is much smaller than the earth, it cooled more rapidly and thus may not have a liquid core, as does the earth, and is not as active internally. Also, the moon, because of its smaller mass, was not able to retain a gaseous atmosphere, and so weathering, erosion, and transfer of surface mass was not possible. About 3 billion years ago, therefore, moonquake and volcanic activity diminished greatly, with perhaps only a vestige remaining today. Since that time the major modifying influence on the moon has been the impact of countless meteors and solar particles.

THE FUTURE promises many important and exciting discoveries from the moon and eventually from other planets of the solar system, through unmanned probes and direct observation and sampling. It is generally true that limited data raise more questions than they answer. Rather than providing an explanation for planetary evolution, therefore, new knowledge from space will result in development of more theories.

MINERAL DESCRIPTIONS

All common and many uncommon mineral species are described and illustrated on pages 70-257. The following explanation of what is and is not included, how it is presented, and why, will enable the reader to use this material to best advantage.

MINERALS INCLUDED in this section were chosen from the 3,000 or more recognized minerals because (1) they are abundant, (2) they have commercial value, or (3) they illustrate an important chemical, structural, or geological fact. Consider the five minerals on page 90. Galena (PbS) is a common and valuable ore of lead. The others are rare. Altaite (PbTe) and clausthalite (PbSe) are included to illustrate that tellurium (Te) and selenium (Se) are so similar, chemically, to sulfur (S) that the two minerals have the same structure as galena, are found in similar geologic environments, and have similar properties. Alabandite (MnS) and oldhamite (CaS) are included because they illustrate that manganese (Mn) and calcium (Ca), in chemical combination with sulfur, form sulfides with the same structure as galena and with similar properties.

The commoner minerals are described in larger type, with relatively detailed information, and are illustrated. Their properties are summarized in smaller type. Rarer species are described in smaller type, with only a few pertinent facts provided, and may or may not be illustrated.

MINERAL ILLUSTRATIONS were planned to provide a good representation of the minerals described. Museum-quality single crystals, when available, have been illustrated to show what real crystals look like, with their irregularities, distortions, and missing faces. For comparison and contrast, idealized crystals are also illustrated to show how the mineral would look without imperfections. The artist's efforts to reproduce the correct color and luster on the faces of the idealized crystals should provide a more realistic picture than line drawings would. A variety of real specimens, including aggregates and mixtures of minerals, has been included to show the diversity of color, shape, habit, and general appearance to be found in minerals.

INFORMATION GIVEN about minerals includes name, chemical formula, important structural features, geologic occurrence and mineral associations, important commercial uses, properties, and any chemical principle the mineral may illustrate. Readers whose interest is sufficiently aroused may wish to pursue the subject further in more detailed and scholarly references (pp. 262-277). Use of this book in conjunction with these references and with real specimens is highly recommended. There is virtually no limit to the information available and the descriptions provided here are to be considered just a beginning.

MINERAL PROPERTIES given in the descriptions are those that are of most help in identifying specimens. An explanation of how the properties are handled is given below. Properties are discussed in detail on earlier pages, as indicated.

- **Crystal system** (p. 38) is given for each mineral, but detailed discussion of the mineral's crystallographic forms is not usually possible within the available space.

- **Crystal habit** (p. 46) describes the overall appearance of the mineral as found in nature, using such terms as tabular, prismatic, and octahedral for single crystals, and terms such as granular, massive, and radiating clusters for aggregates of crystals.

- **Cleavage** (p. 52), like the crystallography on which it is based, is difficult to summarize in any detail without a specialized vocabulary, so the presence of cleavage is noted only in terms of the number of directions in which it occurs. Only the well-developed cleavage planes are counted.

- **Fracture** (p. 52) is included only if it is informative or characteristic of the mineral. It is described in such terms as hackly, irregular, conchoidal, etc.

- **Hardness** (p. 54) is given as a number or range of numbers on the Mohs' scale and applies to clean (unweathered) single crystals. The hardness of these is relatively insensitive to chemical variation, at least within the rather broad ranges between the Mohs' numbers. Because hardness is a measure of the resistance to scratching, measured hardness may be misleading because of weathering, the tendency of crystal fragments to break off an aggregate, etc.

- **Specific gravity** (p. 54) is given either as a mean or as a range because few minerals have fixed compositions and many contain inclusions of other materials. When a range is given, keep in mind that aggregates of crystals may have bulk specific gravities outside the range. In any case, the distinction between light, medium, and heavy minerals may be made without difficulty and is of value in identification.

- **Color** (p. 56) is described in terms of the most common colors observed. Other colors are possible. No effort has been made to describe subtle shades or tints exhibited unless they are typical.

- **Luster** (p. 56) is listed if sufficiently distinctive. It may not be listed for each mineral of a single chemical class when all have a similar luster, as in the vitreous luster of silicates. In such cases, the exception to the rule may be noted. Be cautious about using luster as a diagnostic feature because weathering may alter luster and because fine-grained aggregates may have a different luster than a clean crystal surface.

- **Other properties** such as opacity, streak, and solubility in acids are noted when important for identification.

ALTERATIONS of a mineral's properties may occur as the result of weathering or tarnishing on exposure to the atmosphere. The complexity of such alteration is so great that only the very rapid or very characteristic alterations are noted.

CHEMICAL TESTS AND BLOWPIPE ANALYSES are time-honored aids to mineral identification, but require detailed explanation to be useful. Space available in this book is not sufficient to give all such information, so only a few easy diagnostic tests—e.g. the effervescence of calcite in acid—are given.

GEOGRAPHIC OCCURRENCE of the common rock-forming minerals and many others is so widespread that there is little use in listing localities. In such cases, a general statement regarding geologic occurrence—e.g. in regional metamorphic rocks—is given. Less common species may occur in hundreds of localities, but are abundant or are good collectors' items in relatively few localities. In such cases, these localities are given.

IDENTIFICATION TABLES of many types have been devised to provide a systematic method for identifying minerals by considering each of several properties in sequence and thus narrowing the possible choices. The first division may be, for example, based on the luster—metallic or nonmetallic. Once the minerals with the opposite characteristic are eliminated, one can then look at the next division, e.g. cleavage. By successive determination of properties, one can eventually arrive at the correct identity.

The primary difficulty with such tables is that somewhere in the process it is probable that a property cannot be properly evaluated. So, though such schemes have some value, they do not permit identification without exception. A set of tables sufficient to identify more than a few common minerals is, moreover, unwieldy and difficult to use. For these reasons, such tables have not been included in this book. Many standard mineralogy texts have them, and the reader is urged to look at them.

FURTHER STUDY in other sources is highly recommended. The mineral descriptions and geologic facts given in this book are considered to be the bare minimum needed for the reader to grasp the scope and complexity of mineralogy. Volumes have been written on highly specialized subjects dealing with minerals. Short of studying all these subjects in detail, a reader can still gain a great deal of insight into the relationships between chemistry and structure and between structure and properties, as well as the geologic processes by which minerals are formed and changed. A natural outgrowth of mineral collecting is an interest in mineral properties as related to structure, then in the chemical principles that determine the structure and to other aspects of chemistry and science as a whole. Thus one can gain valuable insight into the relationships between the universe and himself.

GETTING FAMILIAR WITH MINERALS

Learning to recognize minerals is somewhat like getting acquainted with people. We recognize the faces, body conformations, expressions, mannerisms, and numerous other features of people and associate these with specific names, all largely unconsciously. We are not aware that a classification process, based on physical characteristics, is taking place. Though minerals are much less complex chemically than people, they occur in many variations. Because most of their properties are so variable, the only practical way of learning to recognize minerals is to study a large variety of each species until the variations are recognizable in the same way as people are.

To study minerals in this way, it is obviously necessary to have access to mineral collections in which the identifications have already been made. Fortunately extensive collections are numerous. Many cities have natural history museums with good collections; many colleges and universities, particularly those with geology or mineralogy departments, have excellent collections; the thousands of amateur mineralogy and "rockhound" clubs have members who have spent years collecting minerals. In general, the curatorial staff of museums, the staffs of geology and mineralogy departments, and amateur collectors are willing to assist the beginner in the identification of specimens and to provide other information. The libraries of the geology departments or science departments of most colleges and universities have compilations of museums, geology and mineralogy departments, geological surveys, and mineralogy clubs. It is only necessary to write or to visit them to obtain information.

NAMES were given to the commoner minerals and some ore minerals many centuries ago. The name quartz, of Germanic origin, is an example; its meaning has been lost. The name feldspar is derived from a German word meaning, roughly, "field rock," in reference to its abundance. As the study of mineralogy developed, the naming of minerals became more formalized. In general, minerals are given names ending in the suffix "-ite," from the Greek **lithos** (rock). The stem of the name may be derived from a word or words referring to some characteristic of the mineral, e.g. orthoclase (straight fracture) and azurite (blue). Many minerals are named for places where they are found, e.g. montmorillonite (Montmorillon, France) and Terlinguaite (Terlingua, Texas), or for notable persons, e.g. goethite, bunsenite, and smithsonite. Some mineral names indicate their chemical compositions, e.g. chromite, beryl, and borax.

The International Mineralogical Association, in cooperation with member societies in many countries, is responsible for the validity of mineral species and their names. Professional mineralogical journals periodically report decisions of the Association regarding new mineral names and new data.

HOW TO USE THIS BOOK

Learning to identify species of trees is relatively simple with the aid of a field guide because the genetic inheritance of the physical characteristics of living organisms is a precise process. A red maple leaf, for example, looks more or less like all other red maple leaves and like no other leaves. Thus the pictures and descriptions in a book about trees can be compared with an actual tree for on-the-spot identification. Though characteristics of trees change somewhat during growth and vary a little among individual trees of the same species, identification is fairly easy. The same is true for birds, mammals, and other living organisms.

Though minerals are vastly simpler, chemically, than living things, the variations in appearance among different specimens of the same mineral are almost overwhelming. It is not really possible, therefore, to use this or any other book about minerals by itself as a **field guide**. A better suggestion is to use this book as a **study guide**. Compare the descriptions and illustrations with known specimens in museums and in other collections. Equally important, use this book with other books on mineralogy and geology. Only after much study and examination of known minerals is it possible to go into the field and make positive identification of minerals. In conjunction with the field identification of minerals, it is essential to observe relationships among minerals in rocks and between the rocks themselves. Interest in minerals will soon develop into an interest in the origins of rocks and geologic processes affecting them.

This book is designed for the use of serious amateur mineralogists and beginning students of all ages. It is intended to fill the gap between existing popular books, with illustrations and descriptions of limited numbers of minerals, and college-level textbooks, with voluminous data but virtually no illustrations. The jump from one to the other, without professional instruction, is nearly impossible. After a quick reading of the introductory section (pp. 4-69), it is instructive to scan the mineral descriptions (pp. 70-257) to gain an appreciation of the variety of minerals described. Working with known mineral specimens, one can read the description and study the illustrations of a particular mineral, comparing these with a specimen of that mineral and noting the variability among different specimens. Soon it will be possible to recognize the mineral as found in the field.

Learning to recognize minerals is only a beginning, however. The real satisfaction in mineralogy is in gaining knowledge of the ways in which minerals are formed in the earth, of the chemistry of the minerals, and of the ways in which atoms are packed together to form crystals. This book can get you started. From here one can go to mineralogy textbooks, to geology books, to mineral collections, and into the field to collect. Only after enormous effort can one really appreciate the beauty of the mineral kingdom and the satisfaction of understanding it.

CLASSIFICATION OF MINERALS

One of the goals of mineralogy, as in any science, is the classification of the objects to be studied. Only by grouping minerals into definite categories is it possible to study, describe, and discuss them in a systematic and intelligent manner.

A good classification scheme must group minerals so that those in each category share a large number of characteristics. Because minerals are complex, naturally occurring objects, they do not fall neatly into categories. There is, therefore, no perfect classification. Every classification scheme, however, is useful for some specific purpose.

Minerals may be grouped, for example, according to crystallographic characteristics. All cubic minerals make up one crystal system, all tetragonal minerals make up a second system, and so on (pp. 38-43). Each system, in turn, is divided into categories called space groups. There are 230 of these groups. This elaborate classification is necessary for the crystallographer, but is of little value to the average mineral collector or to a mineralogist who does not use X-ray techniques in his work.

Another classification is based on the internal structures of minerals and the kinds of chemical bonds between their atoms (p. 20). All crystals in which the bonding is largely ionic are placed in a single category. This is then further divided into all those with the sodium chloride structure in one subcategory, those with the spinel structure in another, etc. A structural classification of this kind is important to the crystal chemist, but is of limited value to others. Structures and chemical compositions are closely related, however, so the structural classification may be used in conjunction with the chemical classification described in the following paragraphs.

The commonest method of classifying minerals is by chemical composition. A few minerals, the simplest ones, contain only one element or a solid solution (p. 32) of two or more elements. These are classified simply as the native elements and are divided further into native metals, native nonmetals, and native semimetals. The majority of minerals, of course, are compounds, and these are made up essentially of two parts: (1) a metal or semimetal, and (2) a nonmetal or a metal-nonmetal combination called a radical. Compounds are classified according to the chief metal present or according to the chief nonmetal or radical present.

Classification of minerals according to the chief metal present is of obvious value. Minerals that contain a given metal are grouped into a single category. For example, the minerals sphalerite (ZnS), willemite (Zn_2SiO_4), smithsonite ($ZnCO_3$), and zincite (ZnO) are considered a group because the chief metal of each is zinc (Zn). All, in fact, are valuable as ores of zinc. But despite being grouped together, they are considerably different in appearance, behave differently when treated chemically, and are found in very different geologic environments.

Classification of minerals according to the nonmetals or the radicals present is the most widely accepted system and the one used in this book. For example, barite ($BaSO_4$), celestite ($SrSO_4$), and anglesite ($PbSO_4$) are classified as sulfates because all contain the sulfate radical (SO_4). They all are very similar in appearance, behave similarly when treated chemically, and occur under similar geologic conditions, even though they contain different metals—Ba (barium), Sr (strontium), and Pb (lead). Sr and Pb commonly substitute for Ba in barite, leading to a mixed crystal. The three minerals are isostructural—that is, have a similar crystal structure. They are therefore placed in the barite group, a division of the sulfates.

The sulfates are one of a number of classes of minerals. Some classes are not described in this book because no minerals of these classes are common enough to be included here. A more elaborate classification, complete with a numbering system, may be found in **Dana's System of Mineralogy**.

CLASSIFICATION USED IN THIS BOOK

NATIVE ELEMENTS—METALS

Of the more than 100 chemical elements that make up matter, only about 20 are found in pure, or uncombined, form in the earth's crust. These native metals, nonmetals, and semimetals are shown in the periodic table at right. The other elements in the earth's crust have combined with one another, forming the stable compounds that account for the vast majority of minerals.

Only the least reactive of metals—those of the gold and platinum groups—occur in significant amounts as native elements. A few others occur in lesser amounts. Metals are easily recognized by their metallic luster and their malleability.

GOLD GROUP includes three elements of striking chemical similarity: gold, silver, and copper. They have identical crystal structures, with an atom located at each corner of a cube and on each of its six faces, every atom being surrounded by 12 identical atoms. This is called cubic close-packed.

GOLD, Au, has long been prized for its beauty, resistance to chemical attack, and workability. Because it occurs as a native metal, has a relatively low melting point (1063° C.), and is malleable, early man easily separated it from rock and cast or hammered it into intricate shapes. Gold serves as a monetary reserve and is used in jewelry, scientific apparatus, dentistry, and photographic processes.

Gold occurs usually as disseminated grains in quartz veins (lodes) with pyrite and other sulfides or as rounded grains or sometimes nuggets in placer deposits. These are accumulations of stream-carried sediments containing gold eroded from veins. Gold may be panned from such deposits by washing away the lighter sediments from a pan. Even in valuable deposits, gold particles may be too small to be seen. Ores containing less than an ounce of gold per ton may be mined profitably, particularly by large dredges that work on the surface. Pyrite (fool's gold) is sometimes mistaken for gold, but unlike gold, is brittle, tarnishes, and has a brown (rather than yellow) streak. Gold, in fact, is the only malleable yellow mineral.

Gold crystallizes in the cubic system, forming octahedral and dodecahedral crystals, often distorted into dendritic or leafy growths. Cubic crystals are rare. A soft metal, (hardness, 2.5-3), gold can be made harder by alloying it with copper, silver, and other metals. Most gold contains some silver. Pure gold is dense, with a specific gravity of 19.3, decreasing to 15.6 as silver content increases. Ores other than gold itself include gold selenides and tellurides. Major gold deposits are in U.S.S.R., South Africa, Australia, New Zealand, Ontario, California, Alaska, Colorado, Nevada, South Dakota, Montana, Arizona and Utah.

PERIODIC TABLE OF THE ELEMENTS
showing those elements which are found in the pure state or alloys as minerals

Nonmetals

												6 C					
															16 S		
			26 Ir			29 Cu							33 As	34 Se		
				45 Rh	46 Pd	47 Ag					50 Sn	51 Sb	52 Te			
	73 Ta		76 Os	77 Ir	78 Pt	79 Au	80 Hg				82 Pb	83 Bi			

Metals

STRUCTURE OF GOLD

OCTAHEDRAL
GOLD CRYSTALS
(Idealized)

Hopper

Arborescent
Gold

Gold, Placer Co.
California

Avon,
Montana

Anchorage, Alaska

GOLD NUGGETS

Gold on Pyrite Crystals
with Quartz and Silver
Transvaal,
South Africa

Gold on Quartz
Nevada Co.,
California

GOLD OBJECTS

Watch

Ring

Coin

Ear Plug
(Mixtec)

SILVER, Ag, like gold, is a prized metal. It is used in jewelry, tableware, coins, scientific apparatus, dentistry, and, as silver bromide, for photographic processes. Though similar in chemistry and identical in structure to gold, it is much more reactive. This is shown by its tendency to tarnish black when in contact with sulfur. Native silver is much less common than native gold, but because silver occurs in argentite (Ag_2S) and other sulfides it is more abundant. Silver is found in the veins of granitic rocks, either igneous or metamorphic. It is recognized by its silver color, malleability, and high specific gravity (10.5, or greater if sample contains gold).

Silver crystallizes in cubic system and has no cleavage. It occurs as scattered grains, masses, wire, and crystals distorted into dendritic growths. Unlike gold, it is soluble in acids. Its hardness is 2.5-3. Mexico is the major producer, mining silver from granite rocks of southern Rocky Mts. Deposits also are mined in West Germany, Australia, Czechoslovakia, Norway, Ontario, Montana, Idaho, Colorado, and Arizona.

COPPER, Cu, has the same crystal structure as gold and silver, and forms a complete solid solution series (p. 32) with gold. It is much more reactive, being easily attacked by acids and sulfur. It occurs as a native metal in only a few places, notably the Keweenaw Peninsula of Michigan. Here copper is found as cavity fillings in old lava flows, as intergranular cement in sandstones, and in veins. Most of the world's copper is obtained from large copper sulfide deposits. It is one of the most important metals because of its high electrical conductivity and low cost. Pure copper is used for electronic equipment, wiring, and plumbing. Many alloys are important.

Copper crystallizes in the cubic system, its properties being like those of silver and gold. It is copper-red, but corrodes to green carbonate in air. Specific gravity of 8.96 varies with impurity content. Silver, lead, bismuth, tin, iron, and antimony are common in solid solution. It is found in England, U.S.S.R., Australia, Bolivia, Chile, Peru, Mexico, Michigan, New Mexico, Massachusetts, Connecticut, and New Jersey.

RELATED METALS

MERCURY, Hg, is liquid at ordinary temperatures and found as silvery liquid droplets associated with cinnabar (HgS, p. 84), notably in Italy, Yugoslavia, West Germany, Spain, Texas, and California. It is uncommon, so is not a major ore. Mercury is also known as quicksilver.

LEAD, Pb, is a very rare native metal, so reactive it usually forms compounds. It is bluish-gray, heavy, and soft (can be scratched with a fingernail). Crystals have been found in Sweden (large ones in Vermland), U.S.S.R., Mexico, New Jersey, and Idaho.

SILVER CRYSTAL
(Idealized)

Wire Silver
Montezuma,
Colo.

Silver Ore
Montezuma,
Colo.

Dendritic Silver
McMillan Mine,
Ariz.

Silver Ore.
Cobalt, Ontario

Silver and Copper
Lake Superior, Mich.

Jewelry

Silverware

Copper Crystal
Lake Superior, Mich.

Copper Nugget
Houghton Co., Mich.

Copper with Carbonate Corrosion,
Mohawk, Mich.

Arborescent Copper
Broken Hill,
New South Wales,
Australia

Copper Crystals
Keweenaw Co., Mich.

**ALLOYS OF
COPPER**

**Bronze
with Tin**

**Brass
with Zinc**

Native Lead
Langban, Sweden

**Mercury Droplets
in Cinnabar**
Trinity Co., Calif.

Coinage
10% copper
90% silver

PLATINUM GROUP includes elements 44, 45, and 46 (ruthenium, rhodium, and palladium) and elements 76, 77, and 78 (osmium, iridium, and platinum). These are among the rarest and most useful of all metals. They are very nonreactive, are good conductors, are dense, and have high melting points. Because of these properties, they are used in making components of electrical equipment, crucibles, foil, and wire for high-temperature research, jewelry, and many other products. Only platinum and palladium are found in nearly pure form. The others are found as natural alloys.

PLATINUM, Pt, occurs as a native metal in association with dark iron-magnesium silicate rocks (olivine, pyroxene, magnetite, chromite), generally as fine grains disseminated through the rock. Less commonly it is found in quartz veins and contact metamorphic rocks (p. 12). Placer-type deposits are the most important source, notably on the eastern flanks of the Urals. Platinum has the same crystal structure as gold. It is particularly effective as a contact catalyst in the manufacture of certain chemicals. Because of its resistance to chemical corrosion, it is used to make bars and cylinders that serve as standard lengths and weights.

Platinum crystallizes in the cubic system, with no cleavage, occurring as small flakes, grains, and nuggets. Crystals are cubes and octahedra, but are uncommon. Platinum is white to gray, metallic, and ductile. It always contains small amounts of iridium, rhodium, and palladium, and sometimes iron, copper, gold, and nickel. Major producers are U.S.S.R., South Africa, and Canada. It is also found in Madagascar, Finland, West Germany, Ireland, Australia, Columbia, Brazil, Peru, North Carolina, and California.

NATURAL ALLOYS of platinum-group metals have names that give their compositions: platiniridium, osmiridium, iridosmine, etc. Thus iridosmine contains iridium and osmium, the first in more quantity. Properties generally fall between those of the members, but hardness is greater than that of either.

PALLADIUM, Pd, is very rare as a mineral, though it is always present in platinum ore. It occurs with platinum in Colombia, Brazil, U.S.S.R., and South Africa. Its properties are similar to platinum's, but unlike platinum, it is soluble in nitric acid. Hardness is 4.5-5, specific gravity 12.16. Varies with Pt content.

OTHER NATIVE METALS

TIN, Sn, is extremely rare as a native metal. Rounded grains have been reported from Oban, Australia, and tin has been present in volcanic gases from the Italian islands of Stromboli and Vulcano. Chief source of tin is the mineral cassiterite (p. 144). Used extensively in alloys.

IRON, Fe, is rarely found as a native metal because it oxidizes readily to hematite (p. 136), limonite (p. 150), or goethite (p. 148). It occurs in basalts in Greenland and in organic-rich sediments in Missouri. It is found alloyed with nickel, especially in meteorites.

PERIODIC TABLE OF THE ELEMENTS
showing the triad elements

			VIII		
	26 Fe	27 Co	28 Ni		
	44 Ru	45 Rh	46 Pd		
	76 Os	77 Ir	78 Pt		

The triad elements are VIII subgroup elements, closing the three transition series, and are, like the noble gases, more stable than other metals in the same periods.

Platinum Flakes
Colombia

Platinum Nuggets
Ural Mts., U.S.S.R.

Platinum Crystal

Palladium
Brazil

Osmiridium
Norwich, Conn.

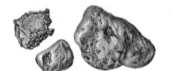

Standard of Length

Crucible

Jewelry

USES OF PLATINUM

Iridosmine
Sisserk, U.S.S.R.

Meteoric Iron
Toluca Valley, Mexico

Meteoric Iron
Taney Co., Mo.

Nickel-iron Alloy (Josephinite),
Terrestrial Mineral
Josephine Co., Ore.

Meteoric Iron
(contains much nickel)
Odessa, Tex.

NATIVE ELEMENTS—NONMETALS

Unlike metals, native nonmetals are brittle and are poor conductors. They form two groups: carbon group and sulfur group.

CARBON GROUP includes diamond and graphite, two forms of solid carbon. The striking differences in their properties are the result of differences in their structures.

DIAMOND, C, forms at high temperatures and pressures in deep volcanic pipes. Originally composed of olivine and phlogopite, the pipes are altered chemically by ground water to a soft blue material, kimberlite, from which the diamonds are easily removed. Erosion carries away many diamonds, which are later recovered from stream deposits. The largest diamond, the Cullinan or Star of Africa, weighed 3025 carats (21 ounces) before cutting. Large stones are rare, however, and most, because of flaws or undesirable color, are used for industrial grinding and cutting. Even though diamond is the hardest known mineral, it is brittle and will cleave readily when struck, and will burn in air at 1,000° C.

Diamond crystallizes in the cubic system, occurring as octahedral crystals, some with curved faces and striations. It has brilliant or greasy luster, is transparent to translucent, and may contain inclusions of graphite, magnetite, and many other minerals. It may be colorless, yellow, brown, black, blue, green, or red, depending on impurities. Hardness is 10, specific gravity 3.5. Commercial deposits are mined in South Africa and Brazil; others in Australia, India, and Arkansas.

GRAPHITE, C, is formed by metamorphism of sedimentary rocks containing organic material or carbonates. Most graphite occurs as small inclusions in metamorphic rock, some large deposits having been mined and depleted. In graphite, layers of strongly bonded carbon atoms are held together by weak van der Waals forces and so are easily moved across one another. Thus it is readily crushed and is a good lubricant. Unlike diamond, graphite does not require high pressures for formation; it can be formed from any organic material heated in an oxygen-poor atmosphere.

Graphite crystallizes in the hexagonal system with perfect cleavage in one direction. It occurs as disseminated flakes, scaly masses, or foliated aggregates. It is black or gray, with metallic, greasy, or dull luster. It commonly contains clay particles. Hardness is 1-2, specific gravity 2.23. Graphite is found in many areas of metamorphic rocks.

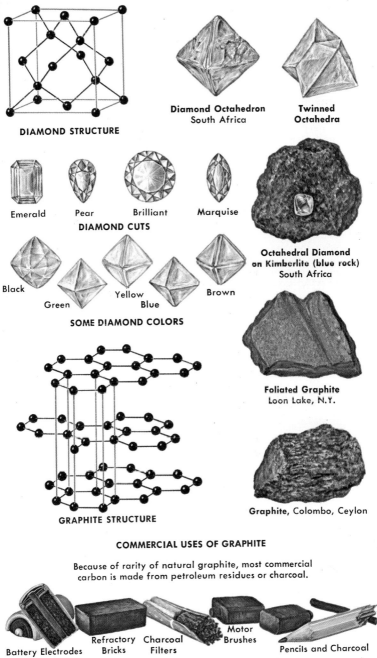

DIAMOND STRUCTURE

Diamond Octahedron
South Africa

Twinned Octahedra

Emerald Pear Brilliant Marquise

DIAMOND CUTS

Octahedral Diamond on Kimberlite (blue rock)
South Africa

Black Green Yellow Blue Brown

SOME DIAMOND COLORS

Foliated Graphite
Loon Lake, N.Y.

GRAPHITE STRUCTURE

Graphite, Colombo, Ceylon

COMMERCIAL USES OF GRAPHITE

Because of rarity of natural graphite, most commercial carbon is made from petroleum residues or charcoal.

Battery Electrodes Refractory Bricks Charcoal Filters Motor Brushes Pencils and Charcoal

SULFUR GROUP consists of three native nonmetals: sulfur, tellurium, and selenium. These nonmetals form crystals made up of ring or chain molecules, the atoms of which form two covalent bonds each (p. 24).

SULFUR, S, exists in three forms. It is almost always found in the form of common sulfur. The other two forms occur very rarely as minerals. The sulfur molecule consists of eight atoms in a puckered ring, the pucker resulting from the atoms being stepped alternately up and down along the ring. The molecules are held together by weak van der Waals forces. Sulfur is commonly precipitated from gases emitted from volcanoes and fumaroles, and is also associated with gypsum (hydrous calcium sulfate), from which it is derived with the aid of organic processes. Sulfur deposits are of relatively recent origin, older deposits having been converted to sulfuric acid. Sulfur melts at 113° C. and burns in oxygen at 270° C., properties that distinguish it from any minerals that may be confused with it. Sulfur is insoluble in water, but is mined by being melted with superheated steam and pumped from underground deposits. It is used in the manufacture of sulfuric acid, rubber, and paper.

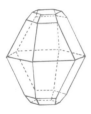

Common sulfur crystallizes in the orthorhombic system (the two rare forms in the monoclinic system). It occurs as blocky crystals, granular masses, or in stalactitic or encrusting form. It is brittle, has resinous or dull luster, and is translucent to transparent, in shades of yellow, brown, or gray. Some tellurium and selenium are commonly present in it. Hardness is 1.5-2.5, specific gravity 2.07. Sulfur is common in volcanic areas, notably in Italy, Sicily, Mexico, and various Pacific islands. Other deposits, associated with salt, occur chiefly in Texas, Louisiana, and the U.S.S.R.

TELLURIUM, Te, is used in manufacturing semiconductors for electronic devices. Most commercial tellurium, however, is refined from tellurides. Tellurium forms a complete solid solution series (p. 32) with selenium and has the same chain molecules as one form of selenium. It crystallizes in the hexagonal system, has a hardness of 2-2.5, is brittle, has a metallic luster, and is opaque and tin-colored. It occurs chiefly in Turkey, Hungary, Australia, and Colorado. **CAUTION: poisonous.** When heated in air, it oxidizes readily to produce toxic fumes, having a garlic odor, that contain TeO_2. Do not breathe the dust.

SELENIUM, Se, has been found only as a sublimate from fires in mining areas. It occurs either with the chainlike structure of tellurium or the ringlike structure of sulfur. It is used in various electronic devices and in some infrared-transmitting glasses. It crystallizes in the hexagonal system, is a metallic gray, and has a red streak. Specific gravity is 4.84. It has been found in West Germany, Arizona, and New Mexico. **CAUTION: selenium is highly poisonous.** When heated in air, it oxidizes readily to produce toxic fumes that contain SeO_2. Do not breathe dust. Keep in a sealed container.

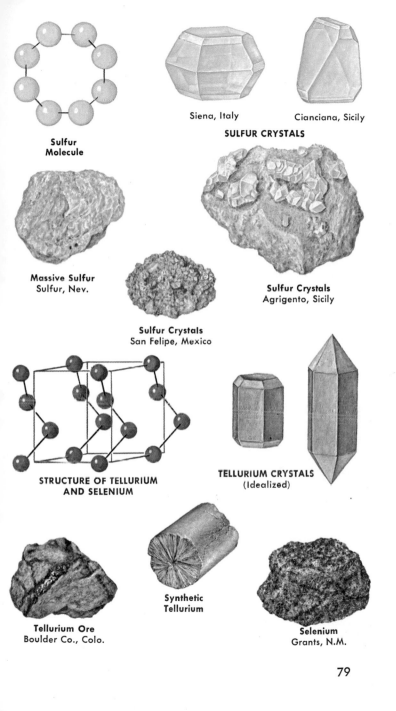

Sulfur Molecule

Siena, Italy

Cianciana, Sicily

SULFUR CRYSTALS

Massive Sulfur
Sulfur, Nev.

Sulfur Crystals
San Felipe, Mexico

Sulfur Crystals
Agrigento, Sicily

STRUCTURE OF TELLURIUM AND SELENIUM

TELLURIUM CRYSTALS
(Idealized)

Tellurium Ore
Boulder Co., Colo.

Synthetic Tellurium

Selenium
Grants, N.M.

NATIVE ELEMENTS—SEMIMETALS

Semimetals fall between metals and nonmetals in the periodic table and exhibit properties of both. Antimony, arsenic, and bismuth are the only native elements generally recognized as semimetals; some authorities also include selenium and tellurium.

ANTIMONY, Sb, occurs in hydrothermal veins with various sulfide ores, but is quite rare. Stibnite, an antimony sulfide (p. 86), is the chief source of the element. Small amounts of antimony are alloyed with lead or other soft metals to strengthen them for use as type metal, battery plates, and cable coverings.

Antimony crystallizes in the rhombohedral system and has perfect cleavage in one direction. It occurs as granular masses or radiating crystal groups. It is brittle, opaque, metallic, and tin-white in color. It has a hardness of 3 and specific gravity of 6.88. It is found in West Germany, France, Sardinia, Sweden, Australia, Chile, Mexico, Canada, and California. **CAUTION: antimony and its compounds are poisonous and should be handled with care.**

ARSENIC, As, is found in hydrothermal veins with other semimetals and with silver, cobalt, and nickel ores. It is less metallic than antimony and bismuth. It is uncommon. Most commercial arsenic is a byproduct of the extraction of metals from arsenides. Arsenic compounds are widely used as insecticides.

Arsenic crystallizes in the rhombohedral system, with perfect cleavage in one direction. It occurs as granular masses or needlelike crystals. It is a metallic white, but tarnishes to dull gray. It volatilizes readily in a flame, with poisonous fumes having a garlic odor. Hardness is 3.5, specific gravity 5.70. Deposits occur in West Germany, France, Chile, Australia, Japan, Canada, Arizona, and Louisiana. **CAUTION: poisonous.**

BISMUTH, Bi, is more metallic, more silvery-white, and much less reactive than arsenic and antimony, and is not poisonous. Native bismuth is scarce. Bismuth obtained from sulfides is used as a low-melting plug for automatic fire-sprinkler systems.

Bismuth crystallizes in rhombohedral system, with perfect cleavage in one direction. It occurs as granular masses and is brittle. Hardness is 2-2.5, specific gravity 9.75. It is found in hydrothermal veins with sulfides in West Germany, France, Norway, Sweden, England, South Africa, Bolivia, Connecticut, South Carolina, Colorado, and elsewhere.

ALLEMONTITE, (As,Sb), is not a compound, but a rare natural alloy with variable composition. Characteristics are intermediate between those of arsenic and antimony. It is found in veins with arsenic, antimony, and sulfides, and in lithium-rich pegmatites. It occurs in France, West Germany, Italy, Sweden, Ontario, and Nevada.

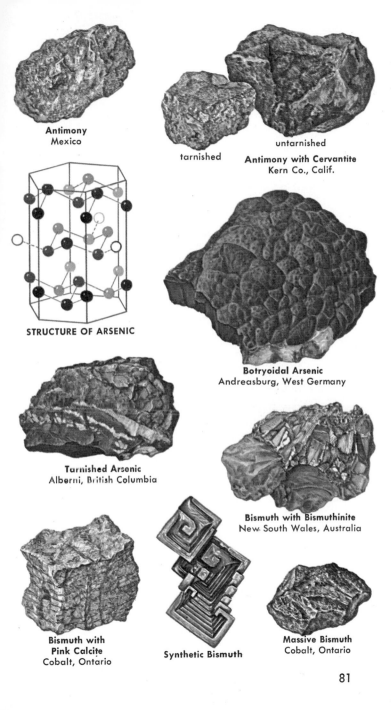

Antimony
Mexico

tarnished

untarnished

Antimony with Cervantite
Kern Co., Calif.

STRUCTURE OF ARSENIC

Botryoidal Arsenic
Andreasburg, West Germany

Tarnished Arsenic
Alberni, British Columbia

Bismuth with Bismuthinite
New South Wales, Australia

**Bismuth with
Pink Calcite**
Cobalt, Ontario

Synthetic Bismuth

Massive Bismuth
Cobalt, Ontario

81

SULFIDES

The sulfides, compounds of sulfur and other elements, are very complex minerals in terms of chemistry and origin. They are, moreover, among the most valuable minerals because they contain many of the metals necessary to civilization. An immense effort has been made to understand the origins of the sulfides and related minerals, but as yet our understanding is not complete, as is the case with many geologic processes.

Sulfides are deposited from aqueous solutions in fracture zones of the crust, located in and near the large igneous bodies called batholiths, in the central cores of major folded mountain ranges. During crystallization of a batholith, its minerals form in a definite sequence, indicated by Bowen's Reaction Series (facing page). Iron, magnesium, and calcium silicates (olivines and calcium feldspars) crystallize first. As the batholith cools, other ferromagnesian minerals and calcium-sodium feldspars crystallize, followed by potassium feldspar, muscovite mica, and quartz. These last three are the major minerals of granite, which makes up the bulk of the batholith we see, after it has been exposed by erosion, millions of years later.

When the granite has crystallized, the remaining material, a small percentage of the total mass, consists of water with sulfur dioxide, sulfur, lead, copper, silver, tin, arsenic, antimony, bismuth, and other transition metals in solution. These are left over because they are unstable at high temperatures and pressures and are soluble in water at lower temperatures and pressures. As the batholith cools and shrinks, fracturing occurs near the margin. Hot-water (**hydrothermal**) solutions move through the fractures to lower-pressure zones nearer the surface. As they move outward into cooler zones, the elements precipitate, the metals and semimetals combining with sulfur and forming veins.

The less-soluble elements crystallize at higher temperatures in or near the batholith (**hypothermal**). More-soluble elements crystalize farther away (**mesothermal**) and very soluble elements crystallize at greater distances, up to tens of miles, from the batholith (**epithermal**). Most veins are thus zoned, with high-temperature zones near the batholith and low-temperature zones farther away.

Because the temperature at any given location falls with time, hydrothermal veins are quite complex, with several generations of mineral deposition visible in any specimen. Sulfides with large amounts of gangue will be deposited; at a later time, at a lower temperature, new minerals will be deposited in small fractures and between the grains. They may completely or partially replace the old minerals. If the old crystal is completely replaced, the new crystal may assume its shape, then being called a **pseudomorph**. If the old crystal is only partly replaced, its remains may be skeletal, completely surrounded by the new crystal. Detailed study is necessary to understand the crystallization sequence.

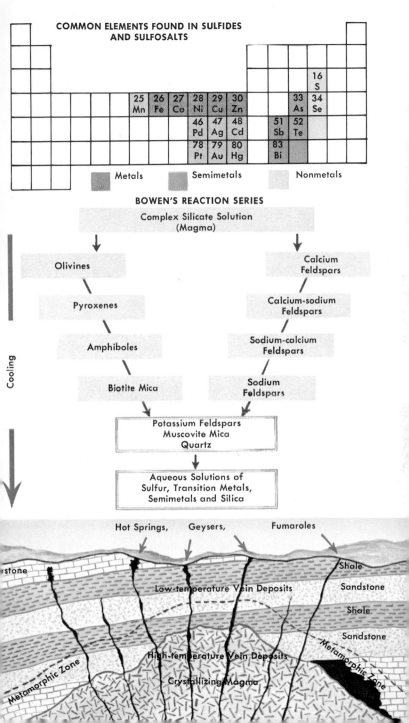

COMMON ELEMENTS FOUND IN SULFIDES AND SULFOSALTS

25 Mn	26 Fe	27 Co	28 Ni	29 Cu	30 Zn		33 As	34 Se	16 S
46 Pd	47 Ag	48 Cd		51 Sb	52 Te
78 Pt	79 Au	80 Hg	83 Bi

Metals Semimetals Nonmetals

BOWEN'S REACTION SERIES

Complex Silicate Solution (Magma)

Olivines

Pyroxenes

Amphiboles

Biotite Mica

Calcium Feldspars

Calcium-sodium Feldspars

Sodium-calcium Feldspars

Sodium Feldspars

Potassium Feldspars
Muscovite Mica
Quartz

Aqueous Solutions of Sulfur, Transition Metals, Semimetals and Silica

Cooling

Hot Springs, Geysers, Fumaroles

Limestone Shale Sandstone Shale Sandstone

Low-temperature Vein Deposits

High-temperature Vein Deposits

Crystallizing Magma

Metamorphic Zone

REALGAR, AsS (arsenic sulfide), differs from other red-orange minerals in having a red-orange streak. A very unstable mineral, it gradually changes to a yellow powder (orpiment) when exposed to light. It occurs in ore veins with orpiment, stibnite, silver, and gold; in volcanic lava, as at Mt. Vesuvius; and in hot-spring deposits, as at Yellowstone National Park.

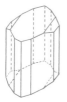

Realgar crystallizes in monoclinic system as granular or crusty masses, rarely as striated stubby prisms, sometimes twinned. It is resinous in luster, translucent to transparent, and sectile. Nitric acid decomposes it. Cleavage is good in one direction, fracture conchoidal. Hardness is 1.5-2, specific gravity 3.59. Occurs in Washington, Rumania, E. and W. Germany, Czechoslovakia, Yugoslavia, Switzerland, etc.

ORPIMENT, As_2S_3 (arsenic trisulfide), is commonly formed by alteration of realgar and other arsenic minerals and is normally found with them in low-temperature hydrothermal veins and in volcanic areas. Its yellow color contrasts with red of realgar. Only other yellow mineral found in similar environment is sulfur, which lacks orpiment's perfect cleavage in one direction and flexible cleavage flakes.

Orpiment crystallizes in monoclinic system as a short, small prisms, rarely distinct, and as granular or foliated masses. It is sectile, resinous in luster, translucent to transparent, and soluble in alkalies. Streak is yellow; hardness 1.5-2; specific gravity 3.48. Occurs in Utah, Nevada, Wyoming, California, Rumania, Germany, Yugoslavia, Switzerland, France, Italy, U.S.S.R., Peru, China, and Japan.

CINNABAR, HgS (mercuric sulfide), is the major ore of mercury. It is found with realgar, orpiment, and stibnite in low-temperature hydrothermal veins and in volcanic deposits. Largest deposit, at Almaden, Spain, occurs as veins in shales and quartzites. It is recognized by its red color, red streak, and heaviness.

Cinnabar crystallizes in rhombohedral system as thick tabular and columnar or prismatic crystals, often twinned, and as granular masses. It is sectile, transparent in thin flakes, brilliant to dull. When it is heated in open tube, mercury droplets form on tube's walls. Hardness is 2-2.5, specific gravity 8.05. Occurs in California, Oregon, Utah, Arkansas, Mexico, Peru, U.S.S.R., China, Yugoslavia, Italy, West Germany, etc.

GREENOCKITE, CdS (cadmium sulfide), crystallizes in hexagonal system usually as coatings on zinc minerals, sometimes as small crystals with a pyramid at one end. It is brittle, yellow to orange, resinous in luster; has complex cleavage, conchoidal fracture. Hardness is 3-3.5, specific gravity 4.77. Found in New Jersey, Missouri, Arkansas, California, Scotland, France, Sardinia, Greece, etc.

Realgar
Ophir District,
Utah

Realgar on Quartz
Manhattan, Nev.

REALGAR CRYSTALS
(Idealized)

**Coarsely Crystalline
Orpiment with
Realgar Veins**
Manhattan, Nev.

**Orpiment with Realgar
in Calcite**
Manhattan, Nev.

**Orpiment and
Realgar**
Manhattan, Nev.

**Cinnabar in
Calcite Veins**
Winnemucca, Nev.

**Cinnabar Crystals
(columnar habit)**
New South Wales,
Australia

**Cinnabar Crystals
in Calcite Gangue**
Terlingua, Tex.

**Greenockite on
Marcasite and Quartz**
Llallagua, Bolivia

Greenockite
Lehigh Co., Pa.

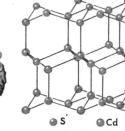

● S   ● Cd
**GREENOCKITE
STRUCTURE**

STIBNITE GROUP includes two sulfides that have similar structures and hence crystals because the elements that form them—antimony and bismuth—are chemically similar. The related mineral guanajuatite is likewise similar because selenium is chemically much like sulfur. Yet these three minerals are found in different deposits and are not equally common.

STIBNITE, Sb_2S_3 (antimony trisulfide), is the most important ore of antimony. It is found in low-temperature veins and hot-spring deposits with realgar, orpiment, and cinnabar. Small deposits are common, large ones rare. Stibnite's crystals, often bent or twisted prisms with an iridescent tarnish, are spectacular and make standard displays in museum collections.

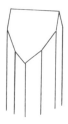

Stibnite crystallizes in orthorhombic system as jumbled aggregates or radiating groups of prismatic crystals or as granular masses. It is an opaque, metallic gray, tarnishing to black, sometimes iridescent. Streak is gray. Cleavage is perfect in one direction, fracture subconchoidal. Hardness is 2, specific gravity 4.63. Stibnite is soluble in hydrochloric acid, is tarnished yellow by potassium hydroxide solution. Occurs in Arkansas, Idaho, Nevada, California, Alaska, Canada, Mexico, Peru, Japan, China, E. and W. Germany, Rumania, Czechoslovakia, Italy, France, England, Algeria, and Kalimantan (Borneo).

BISMUTHINITE, Bi_2S_3 (bismuth trisulfide), is much like stibnite, but is relatively rare and forms in high-temperature veins and in pegmatites. Unlike stibnite, it does not commonly occur in spectacular crystal groups, probably because it does not form in low-temperature, low-pressure environments where crystals have room to grow freely. It normally occurs with bismuth, arsenopyrite, wolframite, and cassiterite.

Bismuthinite crystallizes in orthorhombic system usually as foliated or fibrous masses, seldom as stubby prisms or needles. It is opaque, metallic, gray to white; tarnishes yellow; has gray streak. Cleavage is perfect in one direction. Hardness is 2, specific gravity 6.81. Much selenium replaces S in some samples. Occurs in Utah, Colorado, Montana, Pennsylvania, Connecticut, Canada, Mexico, Bolivia, Australia, England, Sweden, France, Italy, Rumania, and East Germany.

RELATED MINERAL

GUANAJUATITE, Bi_2Se_3 (bismuth selenide), has the same structure as bismuthinite, but with selenium (Se) in the sulfur (S) positions. It is found with bismuthinite in veins near Guanajuato, Mex.; also occurs in Salmon, Idaho; Harz Mts. of West Germany; and Sweden. Characteristics are much like bismuthinite's. Hardness is 2.5-3.5, specific gravity 6.25-6.98. Is soluble in hot aqua regia. **CAUTION: poisonous.**

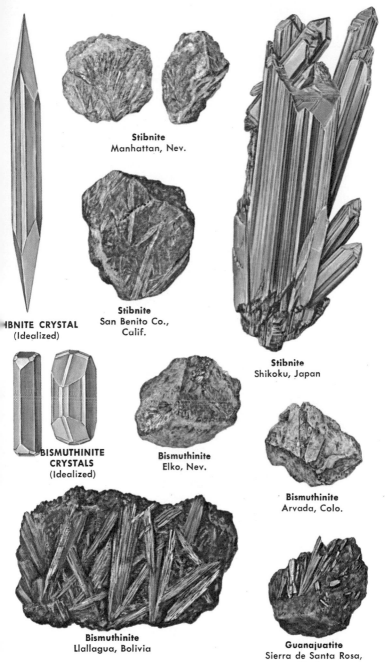

STIBNITE CRYSTAL
(Idealized)

Stibnite
Manhattan, Nev.

Stibnite
San Benito Co.,
Calif.

Stibnite
Shikoku, Japan

**BISMUTHINITE
CRYSTALS**
(Idealized)

Bismuthinite
Elko, Nev.

Bismuthinite
Arvada, Colo.

Bismuthinite
Llallagua, Bolivia

Guanajuatite
Sierra de Santa Rosa,
Mexico

MOLYBDENITE GROUP consists of two sulfides that have a layered structure much like that of graphite, but are heavier than graphite and not as black. The layers are held together by weak van der Waals forces. Within the layers, each molybdenum or tungsten atom is surrounded by six sulfur atoms, at the corners of a right trigonal prism, and each sulfur atom is surrounded by three molybdenum or tungsten atoms. The two minerals are similar in appearance and behavior, but tungstenite is a little harder and much heavier.

MOLYBDENITE, MoS_2 (molybdenum sulfide), is the major ore of molybdenum. It is found in high-temperature veins, pegmatites, and contact metamorphic rocks. Largest source is a deposit at Climax, Colo.

Molybdenite crystallizes in hexagonal system as short tubular prisms, scattered scales, and foliated masses. Layers are flexible. It is opaque, metallic, and gray, with greasy feel and greenish to black streak. It decomposes in nitric acid and is soluble in aqua regia. Cleavage is perfect one-directional, fracture irregular across sheets. Hardness is 1-1.5, specific gravity 4.7. It is found in Maine, New Hampshire, Connecticut, New Jersey, Pennsylvania, Colorado, New Mexico, Utah, Arizona, California, Washington, Alaska, Canada, Norway, Sweden, U.S.S.R., West Germany, England, Scotland, Portugal, Morocco, S. Africa, Australia, China, Japan, Peru, and Mexico.

TUNGSTENITE, WS_2 (tungsten sulfide), is found only at the Emma Mine, near Salt Lake City, Utah, where it is associated with high-temperature sulfides in limestone.

Tungstenite crystallizes in hexagonal system as aggregates of very fine scaly and flexible flakes. It is an opaque metallic-to-dull gray, with gray streak, and has perfect cleavage in one direction. Hardness is 2.5, specific gravity 8.1.

RELATED MINERALS

DYSCRASITE, Ag_3Sb (silver antimonide), is a relatively uncommon mineral. It is found as a vein mineral in silver deposits with other silver minerals and sulfides. It crystallizes in orthorhombic system as foliated or granular masses, sometimes pyramidal crystals. It is sectile, opaque, silver white; tarnishes gray or yellow; has silver streak, complex cleavage, uneven fracture. Hardness is 3.5-4, specific gravity 9.75. Occurs in Nevada, Ontario, Australia, West Germany.

DOMEYKITE, Cu_3As (copper arsenide), is commonly intergrown with algodonite (Cu_6As.) It crystallizes in cubic system as reniform and botryoidal masses. It is sectile, opaque, metallic to dull, and white to gray, tarnishing to yellow or brown. It has white-to-gray streak, no cleavage, uneven fracture. It is soluble in nitric acid. Hardness is 3-3.5, specific gravity 7.2-7.9. Occurs in copper districts of Michigan and Ontario; also in Mexico, Chile, Sweden, England, and West Germany.

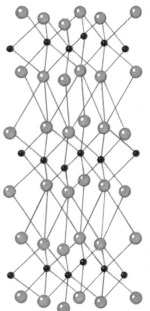

STRUCTURE OF MOLYBDENITE AND TUNGSTENITE

S
Mo or W

Molybdenite with Vein Quartz
Kingsgate,
Australia

Molybdenite
Kingsgate,
Australia

Molybdenite with Quartz in Chlorite
Calaveras Co., Calif.

Molybdenite
Climax, Colo.

Tungstenite
Emma Mine,
Salt Lake City, Utah

Molybdenite with Quartz Gangue
Climax, Colo.

DYSCRASITE CRYSTAL
(Idealized)

Dyscrasite in Quartz
Belmont, Nev.

Domeykite
Houghton, Mich.

GALENA GROUP includes sulfides, tellurides, and selenides that have a halite-type structure (p. 120), with each metallic atom surrounded by six nonmetallic atoms, and vice versa. Galena, altaite, and clausthalite are typical sulfides and illustrate the similarity of the three nonmetals they contain—sulfur (S), tellurium (Te), and selenium (Se). Alabandite and oldhamite are rare because manganese (Mn) normally forms an oxide instead of a sulfide, and calcium (Ca) is more stable in silicate and carbonate structures.

GALENA, PbS (lead sulfide), is one of the most abundant and widely distributed sulfides. The chief ore of lead, it commonly contains valuable amounts of silver. It is found in hydrothermal veins crystallized over a wide temperature range, in contact metamorphic deposits, in large deposits in limestone far from known igneous rocks, in granites, and in lava flows.

Galena crystallizes in cubic system usually as cubes, sometimes as octahedra, often as combinations of both; twin crystals are common. It also occurs frequently as granular masses. It is lead-gray, opaque, metallic in luster, brittle. Streak is gray; cleavage perfect cubic; fracture seldom seen; hardness 2-3; specific gravity 7.57. Nearly always found with sphalerite, commonly with fluorite. May contain zinc, cadmium, copper, arsenic, antimony, other impurities. Occurs in Missouri, Illinois, Iowa, Kansas, Oklahoma, Colorado, Idaho, Utah, Montana, Wisconsin, West Germany, France, Rumania, Czechoslovakia, Austria, Belgium, Italy, Spain, Scotland, England, Australia, Mexico.

ALTAITE, PbTe (lead telluride), is an analog of galena, with Te atoms in S positions. It is found in veins with other tellurides, gold, and sulfides. It is tin-white, has hardness of 3 and specific gravity of 8.27; other properties like galena's. May contain silver, gold, copper, iron. Occurs in Colorado, California, North Carolina, New Mexico, U.S.S.R., Canada, Chile. See p. 259.

CLAUSTHALITE, PbSe (lead selenide), is an analog of galena and thus has similar properties. It occurs in veins and may contain small amounts of mercury, silver, copper, cobalt, and iron. It is lead white and has a hardness of 2.5-3 and a specific gravity of 8.08. Clausthalite occurs in West and East Germany, Spain, Sweden, Argentina, and China. Specimen is illustrated on p. 269.

ALABANDITE, MnS (manganese sulfide), another analog of galena, occurs in low-temperature veins with sulfides and Mn minerals. It is black, with green streak. Hardness is 3.5-4, specific gravity 4.05. Found in Arizona, Colorado, Nevada, Montana, Turkey, Rumania, West Germany, France, Peru, Japan, Mexico. Not highly soluble in water. Illustrated on p. 264.

OLDHAMITE, CaS (calcium sulfide), also has galena structure. It has been found only in meteorites, notably with pyroxenes, at Busti, India, and Bishopville, S.C. It is pale brown, transparent, and soluble in hydrochloric acid and hot water. Cleavage is cubic, hardness 4, specific gravity 2.59. Not found as a mineral on earth because of its solubility.

**Cubic Crystal
with Octahedral Faces**
(Idealized)

**Galena Structure
with Atoms to Scale**

**Elongate Cube
with Octahedral Faces**
(Idealized)

**Twinned Octahedra
of Galena**
(Idealized)

**Octahedral Crystal
with Cube Faces**
(Idealized)

**Galena Cube with Pyrite
and Sphalerite**, Galena, Kan.

**Galena Cube with
Marcasite**, Galena, Kan.

**Perfect Cubic Cleavage
of Galena**

Galena Crystals
Galena, Kansas

Galena Crystals with Pyrite
Joplin, Missouri

Galena in Barite
Kentucky

Galena and Dolomite
Galena, Kan.

91

ARGENTITE GROUP consists of sulfides, selenides, and tellurides of silver (Ag), copper (Cu), and gold (Au). Its members are closely related to the chalcocite group (p. 94) and are major ores of silver and copper. In structure, the argentite minerals are all cubic or become cubic at moderate temperatures. Metallic in appearance, they have similar properties, but differ chiefly in cleavage, brittleness, and color. In addition, their specific gravities vary with their composition. They occur in veins with other silver and copper minerals.

ARGENTITE, Ag_2S (silver sulfide), is chemically identical to acanthite (p. 94), but crystallizes at higher temperatures (above 179° C.) and assumes a cubic structure. Its blackish lead-gray color, high specific gravity (7.04), and ability to be easily cut with a knife help distinguish it. It occurs in low-temperature veins with other silver minerals and notably in microscopic inclusions in galena. It can be synthesized by treating silver with sulfur.

Argentite usually occurs as granular masses or as coatings on other minerals. Crystals are cubic, octahedral, or rarely dodecahedral; they may be distorted in arborescent groups. Streak is gray, cleavage poor, fracture subconchoidal, hardness 2-2.5. Argentite is metallic in luster and opaque. It may contain copper; if rich in copper, it is called jalpaite. It occurs in Nevada, Colorado, Montana, Mexico, Chile, Peru, Bolivia, England, Norway, West Germany, Czechoslovakia.

AGUILARITE, Ag_4SSe (silver sulfide-selenide), is similar to argentite, but Se occupies about half the S positions. Found only at Guanajuato, Mexico, and Virginia City, Nev. It is black, metallic in luster, and shows no cleavage. Hardness is 2.5, specific gravity 7.59. When heated in open tube, it yields metallic silver. **CAUTION: poisonous.**

NAUMANNITE, Ag_2Se (silver selenide), is the selenide analog of argentite. It has been found with clausthalite and other selenides in veins in Idaho, Nevada, Argentina, West Germany. It occurs as cubes, thin plates, and granular masses. It is black, opaque, metallic, malleable. Cleavage is perfect cubic, hardness 2.5, specific gravity 7.87.

HESSITE, Ag_2Te (silver telluride), is the telluride analog of argentite. It is a metallic gray and usually occurs in massive form. It may contain gold. It is found associated with other silver minerals, other tellurides, and gold in small amounts in California, Colorado, Mexico, Chile, Australia, Turkey, U.S.S.R., and Rumania. Streak is gray; cleavage poor; brittle; hardness 2-3; specific gravity 7.88.

PETZITE, Ag_3AuTe_2 (silver gold telluride), is analogous to hessite, with gold in place of 25 percent of the silver. It occurs with hessite and other tellurides in vein deposits in California, Colorado, Ontario, Rumania, and Australia. It is gray to black, with metallic luster, and forms fine granular to compact masses. Cleavage is cubic, fracture subconchoidal. Hardness is 2.5-3, specific gravity 8.7-9.02.

ctahedron with
Cube Faces

Cube

ARGENTITE CRYSTALS
(Idealized)

Argentite
Las Chispas Mine,
Sonora, Mexico

Argentite Crystals
Freiburg, E. Germany

Argentite
Czechoslovakia

Argentite in Calcite
Mexico

Aguilarite
Guanajuato, Mexico

**Naumannite
Crystal**
(Idealized)

Hessite
Hungary

Hessite
Smoky Hill Mine
Boulder Co., Colo.

Petzite
Calaveras Co., Calif.

93

CHALCOCITE GROUP includes minerals that are closely related to the argentite group (p. 92). Chalcocite, stromeyerite, and acanthite are orthorhombic sulfides that are important ore minerals. Digenite and berzelianite are cubic in structure and are often placed in the argentite group. Two minor minerals of the chalcocite group, crookesite and eucairite, are not described here. The extent of solid solution (p. 32) among group members is not certain.

CHALCOCITE, Cu_2S (copper sulfide), is a major ore of copper. It is found most abundantly in the enriched zone of sulfur deposits. It is formed here when percolating water dissolves copper minerals near the surface and redeposits them below the water table, thereby enriching the vein in that zone (p. 103). Chalcocite occurs with bornite and covellite in the enriched zone and with cuprite, malachite, and azurite in the weathered zone above. It is recognized by its dark gray color, its sectility (it is not cut as easily with a knife as argentite, however), and its association with other copper sulfides.

Chalcocite crystallizes in Orthorhombic system (hexagonal above 105° C.), usually as granular masses, rarely as prismatic crystals (sometimes twinned). It is brittle, metallic, opaque, and soluble in acids; alters to malachite, azurite, covellite. Cleavage is prismatic; hardness 2.5-3; specific gravity 5.77. Occurs in Alaska, Arizona, Connecticut, Montana, Utah, Nevada, New Mexico, Tennessee, Mexico, Peru, Chile, U.S.S.R., England, Rumania, and Australia.

ACANTHITE, Ag_2S (silver sulfide), is the orthorhombic form of Ag_2S, argentite (p. 92) being the cubic form. They occur in identical places and have similar properties, most argentite actually changing to acanthite on cooling. Acanthite crystallizes as slender prisms; is sectile, black, metallic, and opaque; has poor cleavage, uneven fracture. Hardness 2-2.5, specific gravity 7.18. See p. 3.

DIGENITE, $Cu_{2-x}S$ (copper sulfide), has a deficiency of metallic (Cu) atoms, a fairly common phenomenon in minerals. It crystallizes in cubic system usually in massive form, sometimes as octahedral crystals. It is opaque, blue-black; has octahedral cleavage. Hardness 2.5-3, specific gravity 5.55-5.71. Occurs in Arizona, Montana, Alaska, Mexico, Sweden, South-West Africa.

STROMEYERITE, AgCuS (silver copper sulfide), occurs in veins with other silver and copper sulfides in Colorado, Montana, Arizona, Canada, Mexico, Australia, Poland, U.S.-S.R., Chile, Peru. It crystallizes in orthorhombic system as prisms, masses; is gray-blue, metallic, brittle, soluble in nitric acid; has no cleavage. Hardness 2.5-3, specific gravity 6.2-6.3. Illustrated on p. 4.

BERZELIANITE, Cu_2Se (copper selenide), is selenide analog of chalcocite. It is relatively rare, occurring with other selenides in veins and in iron ores. It crystallizes in cubic system as scattered grains and as crusts. It is silver-white, but tarnishes; metallic; opaque. Hardness is 2, specific gravity 7.23. Occurs in West Germany, Sweden, and Argentina. Illustrated on p. 4.

A₂X SULFIDES
ORTHORHOMBIC
Chalcocite Cu₂S
Stromeyerite AgCuS
Sternbergite AgFeS
Acanthite Ag₂S

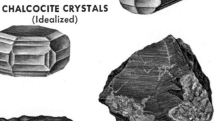

CHALCOCITE CRYSTALS
(Idealized)

Massive Chalcocite
Butte, Mont.

**Chalcocite with Pyrite
and Malachite**
Butte, Mont.

**Chalcocite with
Malachite**
Burra, Australia

**Massive
Chalcocite**
Pima Co., Ariz.

**Chalcocite with
Milky Quartz**
Cornwall,
England

Chalcocite
Butte, Mont.

**Digenite with
Pyrite Flecks**
Bisbee, Ariz.

**Intergrown Berzelianite
and Calcite with Malachite**
Sweden

Berzelianite and Calcite
Sweden

95

SPHALERITE GROUP minerals are similar in structure, and thus in crystal form, cleavage, and behavior. Because they are also chemically similar, they occur in like geologic environments, notably in sulfide veins. Their structure is like that of diamond, with half the carbon positions occupied by sulfur, selenium, or tellurium, and the other positions by zinc, mercury, copper, iron, tin, or some mixture of these metals.

SPHALERITE, ZnS (zinc sulfide), is an abundant sulfide and an important ore of zinc. It occurs in hydrothermal veins in all types of rocks with other sulfides, notably galena. Also, it is commonly a product of contact metamorphism of sedimentary rocks by igneous intrusions (p. 12). Nearly all sphalerite contains much iron in solid solution, which affects its appearance. Its perfect cleavage and resinous luster are diagnostic.

 Sphalerite crystallizes in cubic system as tetrahedral and dodecahedral crystals and as granular and fibrous masses. Twin crystals are common. It is white if pure, more often yellow, brown, black, red, or green; resinous; opaque to transparent; red crystals called semi-ruby; brittle; soluble in hydrochloric acid, with evolution of hydrogen sulfide (rotten eggs odor). Streak is pale yellow to brown; hardness 3.5-4, specific gravity 4.08, varying with iron content. Mineral nearly always occurs with galena. Large deposits are found in limestone of Mississippi Valley, around Joplin, Mo., and Galena, Ill. Also occurs in Colorado, Montana, Wisconsin, Idaho, Ohio, Mexico, Sweden, Britain, Spain, France, Rumania, Switzerland, Czechoslovakia, West Germany, etc.

METACINNABAR, HgS (mercuric sulfide), is an unusual mineral—actually a second form of HgS, more commonly found as the mineral cinnabar (p. 84). It is identical with the black precipitate of HgS that can be formed in the laboratory. When heated to 500° C., metacinnabar will change to the stable form, cinnabar. It is found with cinnabar, other sulfides, and mercury in low-temperature veins. It contains some zinc and selenium. It crystallizes in cubic system as tetrahedral crystals or more commonly as masses. An opaque metallic black, metacinnabar has black streak, hardness of 3, and specific gravity of 7.65. It is found in California, Utah, British Columbia, Italy, Rumania, Czechoslovakia, Spain.

TIEMANNITE, HgSe (mercuric selenide), is a rare mineral that has the sphalerite structure because of the chemical similarity between Hg and Zn and between Se and S. It occurs in Harz Mts., West Germany, and in a limestone vein near Marysvale, Utah. It crystallizes in cubic system usually as metallic gray masses. Hardness is 2.5, specific gravity 8.26. Contains some cadmium and sulfur. **CAUTION: poisonous.**

COLORADOITE, HgTe (mercuric telluride), is found with other tellurides at Kalgoorlie, Australia, and in Boulder Co., Colo. Properties are like those of tiemannite except that it is grayer, has specific gravity of 8.09, and is soluble in nitric acid. **CAUTION: poisonous compound.**

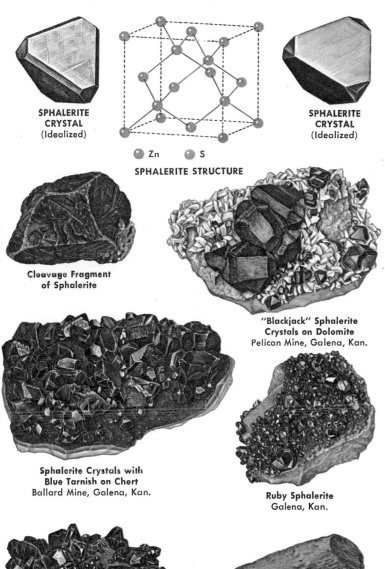

SPHALERITE CRYSTAL
(Idealized)

SPHALERITE CRYSTAL
(Idealized)

● Zn ● S

SPHALERITE STRUCTURE

Cleavage Fragment of Sphalerite

"Blackjack" Sphalerite Crystals on Dolomite
Pelican Mine, Galena, Kan.

Sphalerite Crystals with Blue Tarnish on Chert
Ballard Mine, Galena, Kan.

Ruby Sphalerite
Galena, Kan.

Sphalerite, Semi-Ruby Crystals on Chert
Miami, Oklahoma

Metacinnabar
Idria, Austria

CHALCOPYRITE, CuFeS$_2$ (copper iron sulfide), is the most wide-spread copper mineral and is an important ore of copper. It is commonly found in high- and medium-temperature veins and in contact metamorphic deposits associated with pyrite, cassiterite, or pentlandite and pyrrhotite. It is similar to pyrite (p. 104), but has a more coppery color and a green-black streak.

Chalcopyrite occurs as tetrahedral or sphenoidal crystals, often twins, and as granular or compact masses. It is brittle, brass-yellow (tarnishes darker), metallic in luster, and opaque. It is soluble in nitric acid (yellow S separates). Alters naturally to sulfates, malachite, azurite, and limonite. Cleavage is poor, fracture uneven. Hardness is 3.5-4, specific gravity 4.28. Occurs in New York, Tennessee, Pennsylvania, Missouri, New Mexico, Alaska, Italy, West Germany, France, Spain, Sweden, Norway, Chile, Mexico, Peru, Japan, and elsewhere.

STANNITE, Cu$_2$FeSnS$_4$ (copper iron tin sulfide), has the chalcopyrite structure with half the Fe replaced by Sn. But its gray-black color and blue tarnish distinguish it. It is found in tin-bearing veins with cassiterite, sphalerite, chalcopyrite, and other high-temperature sulfides.

Stannite crystallizes in tetragonal system usually as granular masses or scattered grains, rarely as striated crystals. It is brittle, metallic in luster, opaque; commonly has inclusions of chalcopyrite; may be intergrown with sphalerite; is soluble in nitric acid with separation of Yellow S and black SnO$_2$. Cleavage is poor; fracture uneven, streak black; hardness 4; specific gravity 4.43. Occurs in South Dakota, Alaska Czechoslovakia, England, Australia, Bolivia.

RELATED MINERAL

BORNITE, Cu$_5$FeS$_4$ (copper iron sulfide), is a common copper ore found in high-temperature veins, intrusive igneous rocks, pegmatites, and contact metamorphic rocks, often associated with chalcopyrite and chalcocite. Not in the sphalerite group, it is included here because of its chemical similarity to chalcopyrite. Its copper-red or brown color and purple iridescent tarnish are characteristic and account for its nickname of "peacock ore."

Bornite crystallizes in cubic system rarely as cubic crystals, usually as masses. It is brittle, metallic, opaque; soluble in nitric acid (S separates); alters to chalcocite, malachite, azurite. Streak is gray-black; cleavage poor; hardness 3; specific gravity 5.07. Occurs in Connecticut, Virginia, North Carolina, Colorado, California, Alaska, Austria, West Germany, Italy, England, Malagasy Rep., Australia, etc.

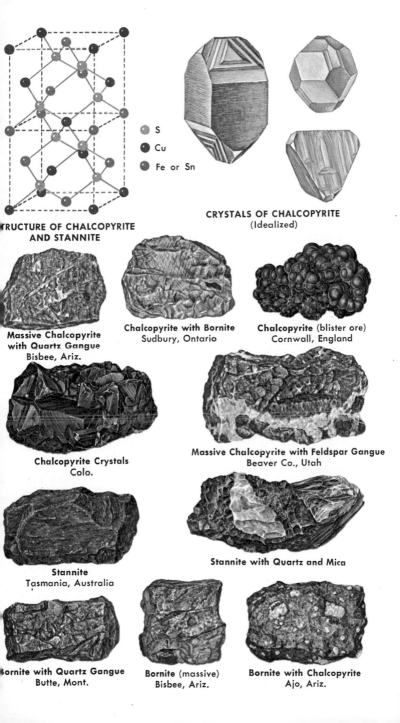

S
Cu
Fe or Sn

STRUCTURE OF CHALCOPYRITE
AND STANNITE

CRYSTALS OF CHALCOPYRITE
(Idealized)

Massive Chalcopyrite
with Quartz Gangue
Bisbee, Ariz.

Chalcopyrite with Bornite
Sudbury, Ontario

Chalcopyrite (blister ore)
Cornwall, England

Chalcopyrite Crystals
Colo.

Massive Chalcopyrite with Feldspar Gangue
Beaver Co., Utah

Stannite
Tasmania, Australia

Stannite with Quartz and Mica

Bornite with Quartz Gangue
Butte, Mont.

Bornite (massive)
Bisbee, Ariz.

Bornite with Chalcopyrite
Ajo, Ariz.

NICCOLITE GROUP includes niccolite, pyrrhotite, and breithauptite. All have a hexagonal structure in which each metal atom is surrounded by six nonmetal atoms, each of which in turn is surrounded by six metal atoms at the corners of a trigonal prism. The closely related minerals covellite, millerite, pentlandite, and klockmannite are included here for convenience.

NICCOLITE, NiAs (nickel arsenide), a minor ore of nickel, is found in basic igneous rocks with other arsenides and with sulfides, or in veins with other high-temperature minerals. Its pale copper-red color is distinctive, but in moist air this may be quickly covered by gray-to-black tarnish.

Niccolite crystallizes in hexagonal system as arborescent or columnar masses, rarely as tabular or pyramidal crystals (sometimes twinned). It is metallic in luster and opaque; has brown-black streak; is intimately intergrown with breithauptite; is soluble in aqua regia; contains some iron cobalt, sulfur. Hardness is 5-5.5, specific gravity 7.83. Occurs in Colorado, New Jersey, Japan, Austria, West Germany, France, Canada, etc.

PYRRHOTITE, $Fe_{1-x}S$ (iron sulfide), is found with pentlandite in basic igneous rocks, veins, and metamorphic rocks. It has a structure similar to niccolite's, but with Fe in some Ni positions and the other positions vacant. The vacancies (x) vary from 0 to 0.2 and cause the mineral to be magnetic, unless their value is near 0, in which case the mineral is nonmagnetic and is called troilite. Bronze color and magnetic properties identify pyrrhotite.

Pyrrhotite crystallizes in hexagonal system as tabular plates and pyramids, often twinned, and as granular masses. Brittle, metallic, and opaque, it has no cleavage and decomposes in hydrochloric acid. Contains some nickel, cobalt, manganese, and copper. Streak is gray-black; hardness 3.5-4.5; specific gravity 4.58-4.79. Occurs in Maine, Connecticut, New Jersey, New York, Pennsylvania, Tennessee, California, Ontario, British Columbia, Rumania, Austria, Italy, West Germany, Switzerland, France, Norway, Sweden, Brazil, Mexico.

PENTLANDITE, $(FeNi)_9S_8$ (iron-nickel sulfide), looks much like pyrrhotite, with which it nearly always occurs. It is commonly intergrown with pyrrhotite, indicating separation (exsolution) from pyrrhotite during cooling. Pentlandite is cubic, however, and is nonmagnetic. It occurs in basic igneous rocks with pyrrhotite, chalcopyrite, cubanite, and other sulfides. The Fe:Ni ration is always 1:1. It can be distinguished from pyrrhotite by its slightly lighter bronze-yellow color, lighter streak, and lack of magnetic character.

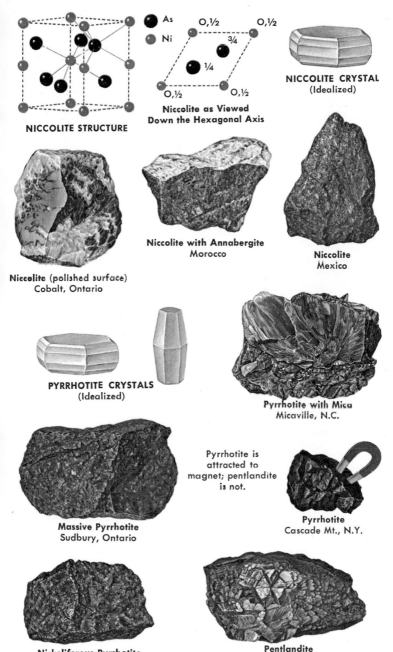

As

Ni

O,½   O,½

3/4

1/4

O,½   O,½

**Niccolite as Viewed
Down the Hexagonal Axis**

NICCOLITE STRUCTURE

NICCOLITE CRYSTAL
(Idealized)

Niccolite with Annabergite
Morocco

Niccolite
Mexico

Niccolite (polished surface)
Cobalt, Ontario

PYRRHOTITE CRYSTALS
(Idealized)

Pyrrhotite with Mica
Micaville, N.C.

Massive Pyrrhotite
Sudbury, Ontario

Pyrrhotite is
attracted to
magnet; pentlandite
is not.

Pyrrhotite
Cascade Mt., N.Y.

Nickeliferous Pyrrhotite
Sudbury, Ontario

Pentlandite
Worthington, Ontario

101

COVELLITE, CuS (copper sulfide), is not abundant, but is important in zones of copper-bearing veins that have been enriched by precipitation from downward-moving water. It is associated and commonly intergrown with chalcopyrite, chalcocite, enargite, and bornite. It is also derived from them. Its indigo or darker blue color, enhanced by wetting, and its yellow and red iridescence are characteristic.

Covellite crystallizes in hexagonal system as thin plates or as massive or spheroidal aggregates. It is opaque and resinous to dull in luster. Streak is gray; cleavage perfect in one direction; hardness 1.5-2; specific gravity 4.60. Contains a little iron. Occurs in Montana, Colorado, Utah, Wyoming, Alaska, Philippines, Yugoslavia, Austria, Italy, Sardinia, East and West Germany, New Zealand, and Argentina.

MILLERITE, NiS (nickel sulfide), is a low-temperature mineral found in veins and cavities of carbonate rocks, in serpentines, in volcanic deposits, and in meteorites. It often forms as an alteration product of other nickel minerals. It is recognized by its brassy color and needlelike crystals.

Millerite crystallizes in hexagonal system as slender to fibrous crystals in radiating or globular groups or tufted coatings. It is brittle; brass yellow, often with gray tarnish; metallic in luster; opaque. Contains some cobalt, iron, copper. Thin crystals are elastic. Cleavage is pyramidal; hardness 3-3.5; specific gravity 5.36. Occurs in Pennsylvania, Iowa, Missouri, Wisconsin, Quebec, Wales, etc.

BREITHAUPTITE, NiSb (nickel antimonide), has the same structure as niccolite (p. 100), with Sb in all As positions. This illustrates the chemical similarity of Sb and As. Extent of solid solution between breithauptite and niccolite is uncertain, but the occurrence of the two minerals together (in calcite veins with silver minerals) may indicate that solid solution is slight. Breithauptite crystallizes in hexagonal system as prismatic or thin tabular crystals or as dendritic growths, disseminated grains, or masses. It is brittle, copper-red with violet tinge, metallic in luster, and opaque. Has no cleavage. Streak is red-brown; brittle; hardness 5.5; specific gravity 8.63. Occurs in Ontario, Sardinia, and West Germany.

KLOCKMANNITE, CuSe (copper selenide), is a rare and not-well-studied mineral that is believed to be similar in structure to covellite. It is impure, and its exact composition is uncertain. It has been found only at Sierra de Umango, Argentina; at Skrikerum, Sweden; and in Harz Mts., West Germany. It occurs with umangite, Cu_3Se_2; eucairite, $CuAgSe$; clausthalite, $PbSe$; and chalcomenite, $CuSeO_3 \cdot 2H_2O$ (an oxidation product of klockmannite). It is believed to crystallize in hexagonal system and occurs as granular aggregates. It is gray, tarnishing blue-black; opaque; metallic in luster. Streak is gray-black; cleavage perfect in one direction; hardness 3; specific gravity about 5. **CAUTION: poisonous.** Do not heat.

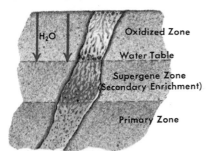

Iron Hat

H₂O

Oxidized Zone

Water Table

Supergene Zone
(Secondary Enrichment)

Primary Zone

Secondary Enrichment of Vein

The primary zone may be too
poor to mine, but movement of
water from the surface dissolves
Cu minerals and deposits them in
the supergene zone, enriching it.

Crystallized Covellite
Butte, Mont.

**Bladed Covellite
with Chert Gangue**

**Massive Covellite
with Chalcopyrite**
Butte, Mont.

**Covellite
with Pyrite**
Butte, Mont.

Millerite with Chalcopyrite
Timagami, Ontario

**BREITHAUPTITE
CRYSTAL**
(Idealized)

**Breithauptite with
Calcite Gangue**
Cobalt, Ontario

**Fibrous Millerite
in Hematite**
Antwerp, N.Y.

PYRITE GROUP consists of very different minerals, but all have a cubic structure. In pyrite, for example, an iron (Fe) atom occupies each corner of a cube and the middle of each face, and a pair of sulfur (S) atoms is midway along each edge. Because the Fe-S bonds in pyrite and similar bonds in other pyrite-group minerals are covalent (p. 24), these minerals are brittle and can be ground to a dull brown powder. They are among the hardest sulfides.

PYRITE, FeS_2 (iron sulfide), is the most abundant and widespread sulfide. It occurs in all types of rocks and veins. Many fossils, particularly in shales, consist of pyrite, formed when hydrogen sulfide from decaying organic matter acted on iron. Often called "fool's gold" because it has been mistaken for gold, pyrite is harder and more brittle. It is harder and yellower than chalcopyrite and yellower than marcasite (besides differing in crystal form). It is used chiefly to produce sulfuric acid. Pyrite, translated freely, means "fire mineral," an allusion to the fact that it gives off sparks when struck. In humid air, it alters readily, forming sulfuric acid and iron sulfate.

Pyrite occurs as cubes, pyritohedra, and octahedra, often twinned, and as granular, radiating, globular, and stalactitic masses. It is pale brass-yellow, tarnishes to brown, and has brown-black streak. It is opaque, metallic in luster, and brittle. Cleavage is indistinct; hardness 6-6.5; specific gravity 5.01. Powder is soluble in concentrated nitric acid. Nickel-containing variety is called bravoite. Important deposits occur in Arizona, Utah, California, Illinois, Virginia, Tennessee, New York, New Hampshire, Connecticut, Pennsylvania, New Jersey, and Ontario.

HAUERITE, MnS_2 (manganese sulfide), is found with gypsum, sulfur, and calcite in low-temperature environments. It is rare. Occurs in salt dome caps in Texas, volcanic rocks in Czechoslovakia, metamorphic rocks in New Zealand, and sediments in Sicily. Red-brown to black, it forms octahedral or cubo-octahedral crystals, globular clusters. Hardness is 4, specific gravity 3.44.

PENROSEITE, $(Ni,Cu,Pb)Se_2$ (nickel-copper-lead selenide), has been found only in Bolivia, associated with pyrite, chalcopyrite, naumannite. Occurs in reniform masses and is gray, opaque, metallic in luster, brittle. Streak is black; cleavage complex; hardness 2.5-3; specific gravity 7.56. Dissolves in nitric acid with effervescence.

SPERRYLITE, $PtAs_2$ (platinum arsenide), is a rare mineral and the only known natural compound of platinum. It is tin-white with black streak, metallic in luster, opaque, and brittle. It forms cubic or cubo-octahedral crystals, often rounded. Cleavage is poor; fracture conchoidal; hardness 6-7; specific gravity 10.59. Occurs in North Carolina, Wyoming, Ontario, S. Africa, U.S.S.R.

LAURITE, RuS_2 (ruthenium sulfide), has been found in platinum placer sands in Borneo and with sperrylite and native platinum in South Africa. It occurs as small crystals or rounded grains and is black, opaque, metallic in luster, and brittle. Streak is gray; cleavage perfect octahedral; hardness 7.5; specific gravity 6.23. Contains some osmium.

PYRITE CRYSTALS
(Idealized)

Pyrite Crystals
Ojinahua Mine,
Chihuahua, Mexico

PYRITE STRUCTURE

S
Fe

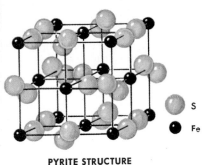

Pyrite
Mexico

Pyrite Crystal
Logroño, Spain

Pyrite and Pyrrhotite
Polk Co., Tex.

Pyrite
Rico, Colo.

Pyrite
Silverton, Colo.

Hauerite Crystals
Raddusa, Sicily

grains   crystal

Sperrylite
Sudbury, Ontario

Banded Pyrite
Binnental, Switzerland

MARCASITE AND RELATED MINERALS described on this page form three groups: (1) marcasite; (2) loellingite, safflorite, and rammelsbergite, which are structurally similar to marcasite; and (3) arsenopyrite, which is structurally different.

MARCASITE, FeS_2 (iron sulfide), has the same formula as pyrite (p. 104), but a different crystal structure. A common mineral, it forms in near-surface deposits where acidity is high, temperature low; in other environments, pyrite forms instead. It is commonest in sediments, particularly clays and lignites. Its peculiar cockscomb habit, when present, distinguishes it from pyrite. It commonly occurs with lead and zinc ores, notably in Joplin, Mo., and Galena, Ill. It deteriorates readily in moist air.

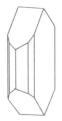

Marcasite crystallizes in orthorhombic system as tabular or prismatic crystals, often with curved faces; as cockscomb aggregates and spearpoint twins; and as stalactitic, globular, concentric, and radiating forms. It is pale bronze-yellow (lighter than pyrite), tin-white on fresh fracture; metallic in luster; opaque; brittle. Streak is black; fracture uneven; hardness 6-6.5; specific gravity 4.88. Found in Kansas, Oklahoma, Wisconsin, Kentucky, Mexico, England, Greece, East and West Germany, Czechoslovakia, etc.

LOELLINGITE, $FeAs_2$ (iron arsenide), is found in medium-temperature veins, often with iron and copper sulfides. May contain cobalt, nickel, antimony, sulfur. It crystallizes in orthorhombic system as prismatic crystals, sometimes twinned, or as massive forms. It is silvery gray; metallic in luster, brittle. Cleavage is prismatic; hardness 5-5.5; specific gravity 7.40-7.58. Occurs in Maine, New Hampshire, New York, New Jersey, Colorado, Ontario, Austria, Poland, Spain, Norway, Chile.

RAMMELSBERGITE, $NiAs_2$ (nickel arsenide), is found with loellingite, skutterudite minerals, sulfides, and other arsenides in medium-temperature veins. It crystallizes probably in orthorhombic system as granular or fibrous masses, rarely crystals. It is tin-white, slightly red; metallic in luster; opaque. Streak is black; hardness 5.5-6; specific gravity about 7.1. Found in East Germany, Austria, France, Italy, Switzerland, Chile, Canada. It alters readily to annabergite.

SAFFLORITE, $(Co,Fe)As_2$ (cobalt-iron arsenide), is cobalt analog of loellingite, normally with 5-16 per cent iron. It crystallizes in orthorhombic system as tabular and prismatic crystals, commonly twinned, and as fibrous masses. It is tin-white (tarnishes gray), metallic in luster, opaque, brittle. Streak is black; cleavage platy; fracture uneven; hardness 4.5-5; specific gravity 7.70. Found in both Germanys, Sweden, Mexico, Ontario. See p. 273.

ARSENOPYRITE, $FeAsS$ (iron arsenide-sulfide), is major ore of arsenic. Occurs in high-temperature veins with silver, copper, and gold ores; in pegmatites; and in contact metamorphic rocks. Crystallizes in monoclinic system as prismatic crystals, columnar and granular masses. It is white-gray, metallic in luster, opaque, brittle. Streak is black; cleavage domal; fracture uneven; hardness 5.5-6; specific gravity 6.18. Widespread occurrence.

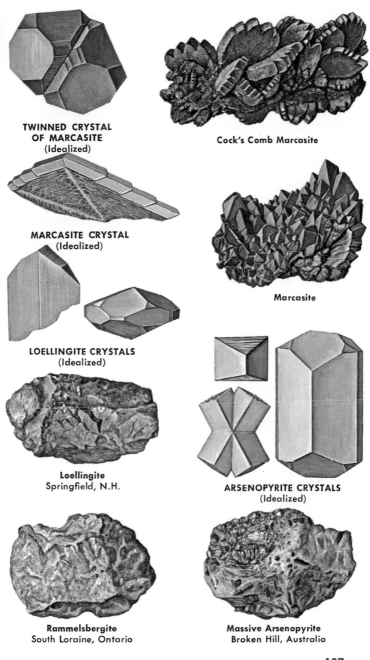

**TWINNED CRYSTAL
OF MARCASITE**
(Idealized)

Cock's Comb Marcasite

MARCASITE CRYSTAL
(Idealized)

Marcasite

LOELLINGITE CRYSTALS
(Idealized)

Loellingite
Springfield, N.H.

ARSENOPYRITE CRYSTALS
(Idealized)

Rammelsbergite
South Loraine, Ontario

Massive Arsenopyrite
Broken Hill, Australia

COBALTITE GROUP includes three minerals with the approximate structure of pyrite (p. 104). Each has an arsenic (As) or antimony (Sb) atom and a sulfur (S) atom bonded in a pair and located in the position of the S pair in pyrite. They are major ores of cobalt (Co). Gersdorffite and ullmannite are sources of nickel (Ni) as well. The Ni, Fe (iron), and Co contents are not fixed, and in ullmannite much Sb and Bi (bismuth) substitute for As. Color and crystal form help distinguish these three minerals from others, but because their appearance is similar and their properties are variable (with composition), they are difficult to distinguish from one another.

COBALTITE, (Co,Fe) AsS (cobalt sulfarsenide), contains up to 10 per cent iron (Fe in formula) and small amounts of nickel. It occurs in metamorphic rocks as disseminated crystals and in igneous rocks as vein deposits with other cobalt and nickel sulfides and arsenides. Complete solid solution (p. 32) probably exists between cobaltite and gersdorffite, but not many intermediate minerals have been found.

Cobaltite crystallizes in cubic system as cubes, pyritohedra, and granular masses. It is silver-white to gray-black with red tinge (color varies with Fe content), metallic in luster, opaque, brittle. Streak is black; cleavage perfect cubic; fracture uneven; hardness 5.5; specific gravity 6.30. Is decomposed by nitric acid with separation of S. Occurs in U.S.S.R., Poland, England, East Germany, India, Australia, etc.

GERSDORFFITE, (Ni,Fe,Co)AsS (nickel sulfarsenide), is the Ni end member of the cobaltite-gersdorffite solid-solution series, with an ideal formula of NiAsS. Many specimens, however, contain up to 16 per cent Fe (iron) and up to 14 per cent Co (cobalt). A variety with up to 13 per cent Sb (antimony) in place of As is called corynite. Gersdorffite is comparatively rare, occurring with other nickel minerals and sulfides in veins. It crystallizes in cubic system as octahedra, pyritohedra, and granular or lamellar masses. It is silver-white to gray, metallic in luster, opaque; decomposes in warm nitric acid with separation of S. Streak is black; cleavage perfect cubic; fracture uneven; hardness 5.5; specific gravity 5.9 (average). Occurs in Connecticut, Pennsylvania, Ontario, Rhodesia, Sweden, Germany, etc.

ULLMANNITE, (Ni,Co) (Sb,As)S, is connected with the cobaltite-gersdorffite series by means of corynite Ni(As,Sb)S, which may be an intermediate member of a gersdorffite-ullmannite solid-solution series. It is essentially isostructural with cobaltite and gersdorffite and is found with the latter and other nickel minerals in hydrothermal vein deposits. Ullmannite crystallizes in cubic system as cubes, octahedra, pyritohedra, and massive forms. Crystals may be twinned. Cube faces are striated. It is gray to white, metallic in luster, and opaque. Streak is black. Hardness is 5-5.5, specific gravity 6.65 (varies with composition). Cleavage is perfect cubic, fracture uneven, brittle. Decomposes in nitric acid. Found in West Germany, Austria, Australia, France, England. Willyamite and Kallilite are varieties of ullmannite.

Tunaberg,
Sweden

Hakansbö, Sweden
(Idealized)

Skutterud, Norway

COBALTITE CRYSTALS

Cobaltite in Slate
Cobalt, Ontario

Cobaltite in Quartz
Cobalt, Ontario

**Cobaltite, High
Grade Ore**
Cobalt, Ontario

Cobaltite
Elliot Lake, Ontario

Gersdorffite
Alora, Spain

Ullmannite
Siegen, West Germany

Corynite
Olsa, U.S.S.R.

Gersdorffite
Salzburg, Austria

Ullmannite
Siegen, West Germany

109

SKUTTERUDITE GROUP, named for the town of Skutterud, Norway, is a complete sold-solution series (p. 32) between smaltite and chloanthite. Any intermediate member is called skutterudite. These minerals have a deficiency of arsenic, indicated by x in the formula; x is 0.5-1 in smaltite and chloanthite, 0-0.5 in intermediate skutterudites. A major source of cobalt and nickel, the minerals are found in medium-temperature veins with other cobalt and nickel minerals. Crystal structure is similar to pyrite's.

Skutterudite minerals crystallize in cubic system as cubic, octahedral, or pyritohedral crystals, commonly misshapen and sometimes twinned, and as netlike forms and granular masses. They are tin-white to gray (tarnish gray), commonly iridescent, metallic in luster, opaque, and brittle. They are soluble in nitric acid, Co-rich members producing red solutions, Ni-rich members green solutions. Streak is black; cleavage indistinct cubic or octahedral; fracture conchoidal or uneven; hardness 5.5-6; specific gravity 6.1-6.9 (depending on composition). Occur in Connecticut, Massachusetts, New Jersey, Colorado, New Mexico, Missouri, Ontario, Norway, both Germanys, Austria, Hungary, Switzerland, Spain, France, British Isles South Africa, Australia, Chile.

SMALTITE, $(Co,Ni)As_{1-x}$ (cobalt-nickel arsenide), is the cobalt-rich skutterudite. It is commonly altered to erythrite, $(Co,Ni)_3(AsO_4)_2 \cdot 8H_2O$, which is red or pink.

CHLOANTHITE, $(Ni,Co)As_{1-x}$ (nickel-cobalt arsenide), is the nickel-rich skutterudite. It is altered to annabergite, $(Ni,Co)_3(AsO_4)_2 \cdot 8H_2O$, the green analog of erythrite.

KRENNERITE GROUP includes calaverite, sylvanite, and krennerite. They are found in low-temperature veins (less commonly in high-temperature veins) with other tellurides, gold, and sulfides. Krennerite has the same composition as calaverite, but crystallizes in the orthorhombic system and is structurally more complex. It is not described here. Sylvanite is probably isostructural with calaverite, but half the metal positions (or less) are occupied by silver. Both crystallize in monoclinic system.

CALAVERITE, Au_2Te_4 (gold telluride), occurs as bladed, striated prisms, commonly twinned, and granular masses. It is brass to silver colored (tarnishes yellow), metallic in luster, opaque, and brittle. Streak is yellow-gray or greenish gray; hardness 2.5-3; specific gravity 9.31. Mineral has no cleavage. It dissolves in hot concentrated nitric acid, releasing metallic Au in red solution. Found in California, Colorado, Ontario, Australia, Philippines. Commonly contains silver.

SYLVANITE, $AgAuTe_4$ (silver gold telluride), occurs as stubby prisms, some twinned, and as skeletal, columnar, or granular forms. It is steel-gray to silver with yellow tinge (streak same), metallic in luster, opaque, brittle. Cleavage is perfect in one direction; fracture uneven; hardness 1.5-2; specific gravity 8.11. Dissolves in nitric acid with separation of metallic Au. Found notably at Cripple Creek, Colo.; also in California, Idaho, Oregon, Ontario, Rumania, Australia.

110

Skutterudite Crystal
Skutterud, Norway

SMALTITE CRYSTALS
(Idealized)

Smaltite
Cobalt, Ontario

Smaltite
West Germany

Chloanthite
Schneeberg, West Germany

Chloanthite
Cobalt, Ontario

Calaverite
Cripple Creek, Colo.

Sylvanite
Cripple Creek, Colo.

111

SULFOSALTS

Sulfosalts are special types of sulfides (p. 82) in which sulfur is combined with one or more metals and one or more semimetals. The semimetals—most commonly arsenic (As) and antimony (Sb)—occupy positions in the crystal that are considered metallic positions. Covellite (CuS), for example, is an ordinary sulfide with Cu (copper) in the metallic positions and S (sulfur) in the nonmetallic. Enargite (Cu_3AsS_4), a sulfosalt, is chemically similar, but the semimetal As occupies a fourth of the metallic positions. This in no way implies that the two minerals are isostructural or that one is derived from the other.

In contrast, pyrite (FeS_2) is a sulfide, but arsenopyrite (FeAsS) is not a sulfosalt. It is a sulfide because the semimetal As occupies nonmetallic rather than metallic positions.

Sulfosalts are geologically similar to sulfides, occurring as vein minerals precipitated from hydrothermal solutions during the last stages of magmatic crystallization. In general they are chemically and crystallographically complex.

RUBY SILVER GROUP includes two deep-red silver minerals. Pyrargyrite is the more common one. There is little evidence of solid solution (p. 32) between the two. They are found together in low-temperature veins with other silver minerals, calcite, and quartz.

PYRARGYRITE, Ag_3SbS_3 (silver antimony sulfide), is a major ore of silver. It is commonly formed by alteration of argentite or native silver and also alters to these minerals.

Pyrargyrite crystallizes in hexagonal system as prismatic crystals or compact masses. It is deep red, brilliant in luster, translucent, brittle. Streak is purple-red; cleavage pyramidal; fracture conchoidal or uneven. Hardness is 2.5, specific gravity 5.82. Decomposes in nitric acid with separation of sulfur (yellow) and antimony trioxide, Sb_2O_3 (white). Found in Colorado, Nevada, Idaho, Ontario, Mexico, Chile, Bolivia, East Germany, Czechoslovakia, Spain.

PROUSTITE, Ag_3AsS_3 (silver arsenic sulfide), is the arsenic analog of pyrargyrite, with As in place of Sb. It may be formed by alteration of argentite or native silver and may be altered to either.

Proustite is like pyrargyrite except for minor differences. It has red streak, may contain some antimony, and decomposes in nitric acid with separation of sulfur (yellow). Hardness is 2-2.5, specific gravity 5.62. It occurs in Colorado, Nevada, Idaho, California, Ontario, Mexico, Chile, East Germany, Czechoslovakia, France, and Sardinia.

Pyrargyrite
Gunnison Co., Colo.

Pyrargyrite
East Germany

Massive Pyrargyrite
Colquechaca, Bolivia

Pyrargyrite and Stephanite
Freiberg, East Germany

Proustite
Chañarcillo, Chile

Proustite
Marienberg, East Germany

Proustite
Mexico

Proustite
Chañarcillo, Chile

TETRAHEDRITE GROUP is a complete solid-solution series of minerals (p. 32) ranging from tetrahedrite, the antimony (Sb) end member, to tennantite, the arsenic (As) end member. Copper is the chief metal in each, but other metals (chiefly iron and zinc) substitute for it extensively. Minerals of this group are among the commonest sulfosalts. Their metallic gray appearance, brittleness, and tetrahedral crystals (when present) help identify them, but they are not easily distinguished without chemical tests.

TETRAHEDRITE, $(Cu,Fe)_{12}Sb_4S_{13}$ (copper-iron antimony sulfide), is one of the commonest sulfosalts. It is found in nearly every major mining area, typically in low- to medium-temperature veins. An important ore of copper, it may contain enough silver to be a valuable source of that metal, too.

Tetrahedrite crystallizes in cubic systems as tetrahedral crystals, often twinned, and granular masses. It is gray-black, metallic, opaque, brittle. It has no cleavage, uneven fracture. Streak is black to brown to red; hardness 3-3.5; specific gravity 4.6-5.1. Found in Idaho, Utah, Montana, Colorado, New Mexico, Nevada, Arizona, California, British Columbia, Bolivia, Peru, Chile, East Germany, Sweden, France, Rumania, Switzerland, Italy, Algeria, England, Austria.

TENNANTITE, $(Cu,Fe)_{12}As_4S_{13}$ (copper-iron arsenic sulfide), is found in low- to medium-temperature veins. Less common than tetrahedrite, it is similar, but is darker, harder (4), and denser (specific gravity 4.6). Occurs in Colorado, Idaho, Utah, Montana, North Carolina, Virginia, Quebec, Ontario, British Columbia, England, Norway, Mexico, Peru, Switzerland, Sweden, East Germany, Poland.

OTHER SULFOSALTS

JAMESONITE, $Pb_4FeSb_6S_{14}$ (lead iron antimony sulfide), is a "feather ore"—i.e. has a feathery or needle-like habit. Relatively uncommon, it forms in low- to medium-temperature veins with other sulfosalts. It is gray-black, is monoclinic in structure, and cleaves in one direction. Hardness is 2.5, specific gravity 5.67. Occurs in many mining areas.

BOULANGERITE, $Pb_5Sb_4S_{11}$ (lead antimony sulfide), a minor ore of lead, is also a feather ore found in veins with sulfosalts and sulfides. It crystallizes in monoclinic system; is blue to gray; oxidizes yellow; dissolves in hydrochloric acid, releasing hydrogen sulfide (rotten eggs odor). Occurs in many areas in small amounts.

ZINKENITE, $Pb_6Sb_{14}S_{27}$ (lead antimony sulfide), is another relatively uncommon feather ore. It crystallizes in hexagonal system, is gray, and has indistinct cleavage. Hardness is 3-3.5, specific gravity 5.22. It is found in low- to medium-temperature veins in regions rich in other sulfosalts. May contain as much as 6% arsenic.

STEPHANITE, Ag_5SbS_4 (silver antimony sulfide), is relatively rare. It crystallizes in orthorhombic system as tabular to short prismatic crystals or massive forms. Color and streak are black; luster metallic; hardness 2-2.5; specific gravity 6.47. It is found in many low-temperature silver veins. Was important silver mineral in Nevada's Comstock Lode.

114

Tetrahedron   Two Tetrahedra   Tristetrahedron   Hextetrahedron

TETRAHEDRITE CRYSTAL FORMS
(Idealized)

**TYPICAL CRYSTAL
OF TETRAHEDRITE**
(Idealized)

TWINNED TETRAHEDRA
(Idealized)

Tetrahedrite with Siderite
Coeur d'Alene, Idaho

**Tetrahedrite with
Quartz Crystals**
Bingham, Utah

Massive Tennantite
Superior, Ariz.

**Jamesonite
with Quartz**
Coeur d'Alene, Idaho

Zinkenite
British Columbia

Fibrous Boulangerite
Stevens Co., Wash.

Massive Boulangerite
Oberhahr, West Germany

Stephanite Crystals

115

ENARGITE GROUP consists of enargite, an arsenic-rich copper sulfide, and famatinite, an antimony-rich copper sulfide. Enargite may contain up to 6 per cent antimony in place of arsenic, and famatinite up to 10 per cent arsenic in place of antimony. The two minerals commonly are intimately intergrown in medium-temperature veins.

ENARGITE, Cu_3AsS_4 (copper arsenic sulfide), is an important copper ore in some areas—at Butte, Mont., for example. Its black color and good prismatic cleavage are keys to its identification.

Enargite crystallizes in orthorhombic system as tabular or prismatic crystals and granular masses. It is brittle, opaque, and metallic in luster (dull if tarnished). Streak is black; hardness 3; specific gravity 4.40. Found in Utah, Nevada, Missouri, Arkansas, Louisiana, Alaska, Mexico, Argentina, Chile, Philippines, Taiwan, Austria, Sardinia, Hungary, Yugoslavia, S.-W. Africa, etc.

FAMATINITE, Cu_3SbS_4 (copper antimony sulfide), is rarer than enargite. Structure is similar to sphalerite's (p. 96). Crystallizes probably in cubic system as minute crystals, granular to dense masses. Is gray tinged with red. Hardness 3.5, specific gravity 4.50. Found in California, Peru, Bolivia, Hungary, Philippines, S.-W. Africa, etc.

BOURNONITE GROUP consists of two lead copper sulfides, one (bournonite) rich in antimony, the other (seligmannite) rich in arsenic. There is incomplete solid solution (p. 32) between them in that bournonite may contain over 3 per cent arsenic in place of antimony. They are found (generally not together) in medium-temperature veins with other sulfosalts and sulfides; also, seligmannite is common in low-temperature environments.

BOURNONITE, $PbCuSbS_3$ (lead copper antimony sulfide), is one of the most abundant sulfosalts and a major ore of lead and copper. It is commonly associated with galena and sphalerite as well as other copper minerals. The Pb:Cu ratio is about 1:1.

Bournonite crystallizes in orthorhombic system as prismatic or tabular crystals and granular masses. It is gray to black (streak same), metallic in luster, opaque, brittle. Cleavage is imperfect; hardness 2.5-3; specific gravity 5.93. Decomposes in nitric acid with white residue. Occurs in many parts of North and South America and Europe.

SELIGMANNITE, $PbCuAsS_3$ (lead copper arsenic sulfide), is relatively rare and not well studied. It is found with other sulfides in cavities in dolomite. It crystallizes in orthorhombic system as tabular or short prismatic crystals, commonly twinned. It is black and brittle. Cleavage is poor; streak brown to purplish black; hardness 3; specific gravity 5.54. Found in Utah, Montana, Australia, Switzerland.

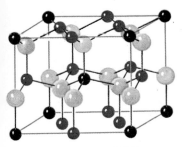

S
Cu
As

ENARGITE CRYSTAL
(Idealized)

ENARGITE STRUCTURE
(note similarity to
Greenockite, p. 85)

**Enargite with Pyrite
and Quartz**
Butte, Mont.

Enargite
National Belle Mine,
Red Mt., San Juan Co.,
Colo.

**Enargite
with Pyrite**
Cerro de Pasco,
Peru

Famatinite
Goldfield, Nev.

Famatinite
Province of
La Rioja,
Argentina

Bournonite Crystals
Herodsfoot Mine,
Cornwall, England

Bournonite Crystals
Germany

**Arsenic-bearing
Bournonite**
Zuni Mine,
Silverton, Colo.

**Seligmannite and
Dufrenoysite**
Switzerland

HALIDES

The halides are salts, the best known being common salt (halite). Each consists of a halogen—fluorine, chlorine, bromine, or iodine —and a metal. The bond between the atoms is highly ionic (p. 22) if the metal is very metallic, or more covalent (p. 24) if the metal is not so metallic.

In general the halides are soluble in water. Even those not soluble at atmospheric temperature and pressure are more soluble than most compounds at higher temperatures and pressures. The halides are therefore found in geologic environments formed at low temperature—in near-surface deposits of hydrothermal origin, near hot springs and geysers, and most notably in evaporite deposits.

Since the earth formed, large amounts of dissolved halides have entered the ocean, making it quite salty. Nevertheless ocean water is not saturated because halides are so highly soluble. When an arm of the sea is cut off from the open water, however, evaporation of water concentrates the salt, which is then deposited in layers on the bottom. This has happened repeatedly in the past, building up thick, extensive deposits of halite and other halides, which are now being mined. Freshwater lakes with no natural outlets, such as Great Salt Lake, become salty in time because, as water evaporates, dissolved salts remain behind, becoming ever more concentrated.

Four uncommon halides are described below. All except calomel have the sphalerite structure (p. 96).

CALOMEL, HgCl (mercury chloride), crystallizes in tetragonal system, usually forming tabular prisms and drusy or earthy masses. It is transparent; colorless, gray, yellow, or brown; adamantine in luster; sectile; and insoluble in water. Hardness is 1.5, specific gravity 7.23. It occurs in West Germany, Yugoslavia, Spain, France, Mexico, Texas, Arkansas, California. Formed by alteration of other mercury minerals.

MARSHITE, CuI (copper iodide), is one of the few natural iodides. It crystallizes in cubic system, forming tetrahedral crystals; is transparent, colorless or pale yellow (weathers to dark red, has white streak), adamantine in luster, insoluble in water, brittle. Hardness 2.5, specific gravity 5.60. Forms complete solid solution with miersite; found in Chile, Australia. Can be formed by action of hydrogen iodide on Cu.

NANTOKITE, CuCl (copper chloride), crystallizes in cubic system usually as granular masses. It is transparent; colorless, green, or gray (streak is white); adamantine in luster; brittle. It decomposes in water, gives off chlorine odor when crushed, dissolves in hydrochloric acid, nitric acid, and ammonium hydroxide. Hardness is 2.5, specific gravity 4.22. It has been found at Nantoko, Chile, and in Australia.

MIERSITE, AgI (silver iodide), crystallizes in cubic system, forming tetrahedral crystals. It is yellow (with yellow streak), transparent, adamantine in luster, brittle. (A hexagonal form is called iodyrite). Hardness is 2.5, specific gravity 5.67 Miersite and marshite form a solid solution series and are reduced to silver and copper by sulfuric acid and zinc. Miersite is found at Broken Hill, Australia.

PERIODIC TABLE OF THE ELEMENTS
showing the Halogens and Metals found in natural Halides

3 Li																	9 F
11 Na	12 Mg											13 Al	14 Si				17 Cl
19 K	20 Ca			25 Mn	26 Fe			29 Cu									35 Br
		39 Y						47 Ag									53 I
55 Cs	57 La									80 Hg		82 Pb	83 Bi				85 At

The halogens, fluorine, chlorine, bromine, iodine and astatine, are the Group VII elements, requiring only one electron to acquire the stable octet. They all ionize readily to -1 ions and bond readily with metal ions. Only those which form halides insoluble in water are found as minerals.

F^{-1} 1.36

Cl^{-1} 1.81

Br^{-1} 1.95

I^{-1} 2.16

Sizes of Halide Ions, Angstroms

CALOMEL CRYSTAL
(Idealized)

Calomel
Bavaria,
West Germany

Calomel with Mercury
Terlingua, Tex.

Nantokite

MARSHITE CRYSTAL
(Idealized)

Marshite
Broken Hill, Australia

Marshite
Broken Hill, Australia

119

HALITE GROUP includes halides with the structure of halite. In halite, NaCl, the Na⁺ ion is so large that it must be surrounded by six Cl⁻ ions to shield it from other positive ions. Each Cl⁻ is likewise surrounded by six Na⁺, as required by both having one charge. The resulting structure is cubic, with identical atoms at the corners of a cube and at the centers of all six faces. In all three axial directions there are rows of alternating Na⁺ and Cl⁻ ions. The halides of Na⁺ and K⁺ (potassium) are very ionic in character (p. 22) and are soluble in water. The halides of Ag⁺ (silver) are considerably more covalent (p. 24) and are insoluble.

HALITE, NaCl (sodium chloride), is the most abundant halide and one of the most thoroughly investigated crystals. It occurs in large deposits, in extensive beds, and in salt domes that have risen from deep beneath the surface. As common table salt, it is one of the most familiar minerals. Its property of enhancing the natural flavor of food makes it a prized substance, and its distinctive taste is an absolute test for its identification. Halite is the major source of chlorine for the manufacture of sodium hypochlorite, used as a bleach and as a disinfectant.

Halite crystallizes in cubic system, has perfect cubic cleavage, and occurs as masses of interlocking crystals—commonly cubes (often with hopper faces), rarely octahedra. It is normally gray, with included clays, sometimes white, yellow, red, blue, or purple; its streak is white. It is transparent, vitreous in luster, and brittle. Hardness is 2, specific gravity 2.17. It is found in U.S.S.R. (famed Siberian salt mines), Austria, West Germany, Poland, England, Switzerland, France, Sicily, Spain, South West Africa, India, Algeria, Ethiopia, China, Peru, Colombia, Ontario, New York, Ohio, Michigan, Texas and Louisiana (salt domes), New Mexico, Utah, California, Kansas, Nevada, and Arizona.

SYLVITE, KCl (potassium chloride), is a salt that occurs in the same manner as halite and is often found with it, but is much less common. This is probably because much of the potassium ions of the earth's crust are retained by the clay minerals rather than being dissolved in water. Though sylvite is very similar in appearance to halite, it is less dense and more bitter in its taste. It is also less brittle, deforming readily under directed pressure.

Sylvite crystallizes in the cubic system, has perfect cubic cleavage, and forms cubic and octahedral crystals, both rare. It usually occurs as granular masses. Vitreous in luster, it is a transparent white, gray, blue, red, or yellow, and has a white streak. Hardness is 2, specific gravity 1.99. It is found in salt basins and in fumaroles in East Germany, Poland, Sicily, Italy, Chile, Peru, New Mexico, Texas.

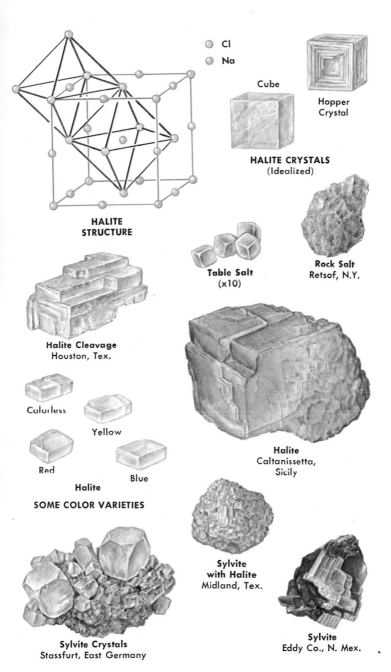

Cl

Na

Cube

Hopper Crystal

HALITE CRYSTALS
(Idealized)

HALITE STRUCTURE

Rock Salt
Retsof, N.Y.

Table Salt
(x10)

Halite Cleavage
Houston, Tex.

Colorless

Yellow

Red

Blue

Halite

SOME COLOR VARIETIES

Halite
Caltanissetta, Sicily

Sylvite
with Halite
Midland, Tex.

Sylvite Crystals
Stassfurt, East Germany

Sylvite
Eddy Co., N. Mex.

121

VILLIAUMITE, NaF (sodium fluoride), occurs in insignificant amounts, but is interesting because it fills small cavities in sodium-rich rocks in Guinea and on the Kola Peninsula of the U.S.S.R. It has the halite structure (p. 120), but its solubility is relatively low. Synthetic villiaumite is colorless, but the mineral is deep red because of defects in its structure. Heating causes the color to disappear.

Villiaumite crystallizes in cubic system, forming cubic crystals (often hoppers) and, more commonly, granular masses. It has perfect cubic cleavage and white streak. It is brittle, glassy, and soluble in hydrofluoric acid. Hardness is 2-2.5, specific gravity is 2.81.

CERARGYRITE, AgCl (silver chloride), often called horn silver, is the chloride end member of a complete solid-solution series (p. 32) with bromyrite, AgBr. Most specimens contain Br; those with more Cl than Br are cerargyrite. Iodine may also be present. A secondary mineral, cerargyrite results from weathering of silver vein minerals, particularly in arid regions where no streams carry away Cl or Br. It is recognized by its hornlike look and its ability to be cut with a knife.

Cerargyrite crystallizes in cubic system, forming cubic crystals (rare) and waxy masses, crusts, and columns. It has no cleavage. White if pure, it is usually gray, changing to purple or brown on exposure to light; bromine browns, and iodine deepens color. Hardness is 2.5, specific gravity 5.55, increasing with bromine content. Insoluble in water, it dissolves in ammonium hydroxide. Chief sources: Australia, Chile, Peru, Bolivia, Mexico, Colorado, California, Nevada, and Idaho.

BROMYRITE, AgBr (silver bromide), is less common than cerargyrite, AgCl, with which it forms a complete solid-solution series. Bromyrite includes all specimens with more Br than Cl. As is typical with such series, the two end members never occur together, but are found in the same type environment (see cerargyrite). Bromyrite crystallizes in cubic system, forming cubic crystals and, more commonly, granular masses. It has no cleavage and is plastic. Color and streak are brown, luster resinous, hardness 2.5, specific gravity 6.5 (for pure AgBr, less with Cl.) Found in West Germany, France, U.S.S.R., Australia, Chile, Mexico, and Arizona.

SAL AMMONIAC, NH₄Cl (ammonium chloride), does not have the halite structure (p. 120) at ordinary temperatures—rather eight Cl⁻ ions surround the (NH₄)⁺ ion—but is included here because its structure changes to the halite type of 194.3° C. It forms as a sublimate near volcanic fumaroles and in guano deposits. It crystallizes in cubic system, has imperfect octahedral cleavage, and occurs as crystals, dendritic shapes, and fibrous and earthy masses. It is gray, white, yellow, or brown, with glassy luster and salty, burning taste, and is brittle. Hardness is 2, specific gravity 1.54. Found in France, West Germany, England, Italy, Peru, Chile.

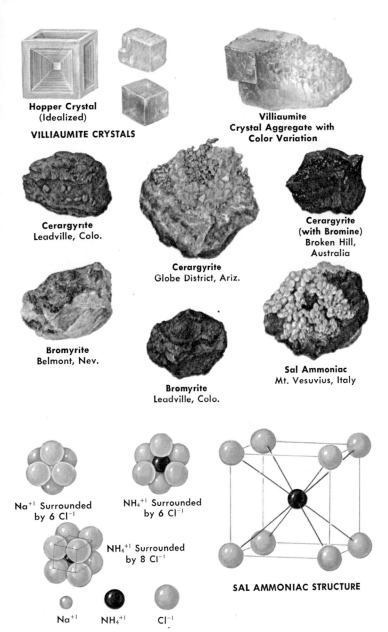

Hopper Crystal
(Idealized)

VILLIAUMITE CRYSTALS

Villiaumite
Crystal Aggregate with
Color Variation

Cerargyrite
Leadville, Colo.

Cerargyrite
Globe District, Ariz.

Cerargyrite
(with Bromine)
Broken Hill,
Australia

Bromyrite
Belmont, Nev.

Bromyrite
Leadville, Colo.

Sal Ammoniac
Mt. Vesuvius, Italy

Na^{+1} Surrounded
by 6 Cl^{-1}

NH_4^{+1} Surrounded
by 6 Cl^{-1}

NH_4^{+1} Surrounded
by 8 Cl^{-1}

Na^{+1} NH_4^{+1} Cl^{-1}

SAL AMMONIAC STRUCTURE

Na^{+1} is stable when surrounded by 6 Cl^{-1} because it is
shielded from other Na^{+1} ions. $(NH_4)^{+1}$ is too large to
be shielded by 6 Cl^{-1} ions. It must be surrounded by
8 Cl^{-1} ions. The structure of Sal Ammoniac is the
same as that of CsCl.

123

FLUORITE GROUP includes fluorite, CaF_2, and other minerals with various amounts of metals less common than calcium, notably yttrium and cesium, in a portion of the metallic positions. The calcium ion, Ca^{++}, because of its double charge, must combine with two fluorine ions, F^-, in a crystal. Because eight F ions are required to shield the positive metal (Ca) ions from other metal ions, the fluorite structure, with each Ca surrounded by eight F and each F surrounded by four Ca, is assumed.

FLUORITE, CaF_2 (calcium fluoride), is found in a wide range of geologic environments, as late-stage crystals in granites, as a gangue material in high-temperature veins, as massive replacement deposits in limestones, and as cavity fillings and cement in sandstones. It can be nearly any color, depending on impurities present, exposure to radioactivity, and temperature of crystallization. Its perfect octahedral cleavage, a result of its structure, and its glassy luster are diagnostic. It is found in large deposits in many areas. Important as a flux in the production of steel and ceramic glazes, it is also used in making high-quality optical lenses for specialized applications. Very large crystals and aggregates are found in northern Ohio, northwestern Illinois, southwestern Missouri, and northwestern Oklahoma, where it is associated with celestite or with galena and sphalerite.

Fluorite crystallizes in cubic system, forming cubes (common) and octahedra (less common), often in twins. It also occurs as granular, earthy, fibrous, and globular masses. It is brittle, glassy (translucent to transparent), and colorless if pure, otherwise amber, purple, blue, green, black, gray, yellow, or less commonly red or pink. Streak is generally white. It forms limited solid solution with yttrofluorite. Only slightly soluble in water, it is decomposed by sulfuric acid. It occurs in E. & W. Germany, Austria, Switzerland, England, Norway, Mexico, Ontario, Illinois, Kentucky, Colorado, New Mexico, Arizona, Ohio, New Hampshire, New York, and many other areas.

YTTROFLUORITE, $(Ca,Y)F_2$—is the name applied to any fluorite with an appreciable amount of Y in place of Ca in the metallic positions. Its properties are very similar to those of pure CaF_2 (fluorite). Yttrofluorite has been found in limestones in New Jersey and New York, in pegmatites in Maine, Colorado, and Massachusetts, and in veins in several Canadian areas. Yttrofluorite is not widely accepted as a valid mineral species, but it is useful to denote a variety of fluorite, with yttrium, Y^{+3}, in the structure.

YTTROCERITE, $(Ca,Y,Ce)F_2$, is a relatively rare fluorite with an appreciable amount of both Ce and Y in Ca positions. Ca, however, is much more abundant than Y and Ce combined in all reported specimens. Yttrocerite is not generally accepted as a valid mineral species, but, as in the case with yttrofluorite, the name is useful for denoting a particular compositional variety of fluorite. Properties are very similar to those of pure fluorite. Yttrocerite has been reported from Sweden, the U.S.S.R., and New York.

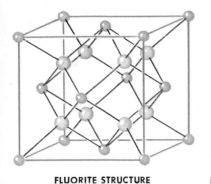

FLUORITE STRUCTURE

FLUORITE CRYSTALS
(Idealized)

○ Ca
○ F

**PENETRATION TWIN
OF FLUORITE**
(Idealized)

Arizona

Rosiclare, Ill.   Arizona

SOME COLOR VARIETIES OF FLUORITE

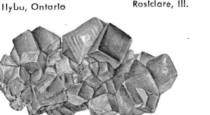

Fluorite in Calcite
Hybla, Ontario

**Octahedral Cleavage,
Fragment of Fluorite**
Rosiclare, Ill.

**Fluorite with
Celestite**
Clay Center,
Ohio

Fluorite
Silverton,
Colo.

Fluorite, Crystal Aggregate
Durham, England

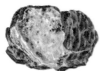

**Yttrocerite and
Yttrocalcite with Mica**
Orange Co., N.Y.

Yttrofluorite
Hundholmen,
Norway

Fluorite
Madoc, Ontario

CRYOLITE, Na_3AlF_6 (sodium-aluminum fluoride), was named after the Greek words for "frost stone" because of its often icy look. It occurs in pegmatites with other fluorine-bearing minerals, particularly topaz; with microcline and quartz, the major pegmatite minerals; with galena, molybdenite, and other sulfides; and with siderite, cassiterite, wolframite, and others. Molten cryolite is used in electrolytic reduction of bauxite to aluminum. Because its index of refraction is nearly the same as water's, it is almost invisible when ground and immersed in water. The structure is made up of $(AlF_6)^{-3}$ octahedra, with Na^{+1} between them.

Cryolite crystallizes in monoclinic (pseudo cubic) system, normally forming granular masses, rarely cubic crystals (commonly twinned). It is colorless, white (if granular), brown, black, or red, with glassy or greasy luster and white streak. Fine granular specimens look like ice. Brittle, it fractures unevenly. Hardness is 2.5, specific gravity 2.96. Cryolite dissolves in sulfuric acid, giving off hydrofluoric acid fumes, and in aluminum chloride solution. Only large deposit is at Ivigtut, Greenland; smaller ones in U.S.S.R., Spain, and Colorado (Pike's Peak).

CARNALLITE, $KMgCl_3 \cdot 6H_2O$ (potassium-magnesium hydrous chloride), loses water at relatively low temperature, which destroys its structure. This and lack of cleavage distinguish it from halite and other salts with which it occurs. It crystallizes in orthorhombic system, forming granular masses. Tabular or barrel-shaped crystals are rare. It is white, seldom yellow or blue, with dull, greasy luster, white streak, and bitter taste. It dissolves in water. Hardness is 2.5, specific gravity 1.60. Occurs chiefly in E. & W. Germany, Spain, Texas, and New Mexico.

ATACAMITE, $Cu_2(OH)_3Cl$ (copper hydroxychloride), has the hydroxyl ion, $(OH)^-$, in its structure. It is formed by alteration of copper minerals in arid climates. It crystallizes in orthorhombic system as slender or tabular prisms, fibrous masses, and sand. Transparent, bright green, with glassy to brilliant luster and light green streak, it is brittle and soluble in acids. Cleavage is complex, hardness 3-3.5, specific gravity 3.76. Named after Atacama Desert, Chile, where first found. Also occurs in Peru, Mexico, Australia, Italy, U.S.S.R., England, and Arizona.

TERLINGUAITE, Hg_2OCl (mercuric oxychloride), like eglestonite, is an alteration product of other mercury minerals, but is more highly oxidized and has both mercury ions, Hg^+ and Hg^{+2}. It has been found only in mercury deposits at Terlingua, Tex. It crystallizes in monoclinic system, forming tabular prisms and powdery aggregates. It is lemon yellow to greenish, tarnishing to olive green, brilliant in luster, brittle, soluble in acids. Hardness 2.5, specific gravity 8.73.

EGLESTONITE, Hg_4OCl_2 (mercuric oxychloride), is an alteration product of mercury minerals, less highly oxidized than terlinguaite; only Hg^+ is present. It occurs as dodecahedral crystals, masses, and crusts in cubic system. It is translucent yellow-orange, tarnishes brown to black, and streaks yellow to greenish yellow. It is decomposed by acids to calomel, $HgCl$. Hardness is 2.5, specific gravity 8.61. Found at Terlingua, Tex.; in Pike Co., Ark.; near Palo Alto, Calif.; and in South Africa.

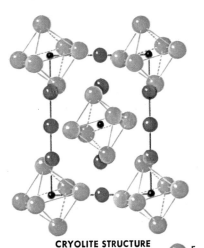

CRYOLITE STRUCTURE

F
Na
Al

Pure Cryolite
Ivigtut, Greenland

**Cryolite with
Sphalerite**
Ivigtut, Greenland

Carnallite
Moab, Utah

Atacamite
Wallaroo, Australia

Atacamite
Alacama Desert,
Chile

Carnallite
Carlsbad, N. Mex.

Atacamite
Los Remolines,
Chile

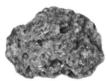

**Terlinguaite and
Eglestonite with
Mercury**
Terlingua, Tex.

OXIDES AND HYDROXIDES

Oxides are compounds that combine metals or semimetals and oxygen. The simple oxides have only one metal, which combines with oxygen in the proportion necessary to balance the charges on their respective ions, though in some cases the metal-oxygen bonds are highly covalent (p. 24). Thus Cu^+ (copper ion), Mg^{+2} (magnesium ion), Al^{+3} (aluminum ion), and Si^{+4} (silicon ion) combine with O^{-2} (oxygen ion) and form Cu_2O (cuprite), MgO (periclase), Al_2O_3 (corundum), and SiO_2 (silica). Mixed oxides have more than one metal—for example, $MgAl_2O_4$ (spinel) and $FeTiO_3$ (ilmenite). They are analogous to the simple oxides, but generally have a more complex structure.

Aside from ice, a unique mineral, the most abundant oxide is silica. Silica not only is plentiful, but also combines readily with other oxides at high temperatures beneath the surface of the earth. As a result, simple oxides are rare, and the silicates (consisting of silicon, oxygen, and one or more metals) are the most abundant minerals. The crystalline form of pure silica, quartz, is common; it represents the silica left after all other oxides have combined to form silicates. The other oxides are common only in geologic environments poor in silica. There, they represent the metals remaining after all the silica has been crystallized in the silicates.

The hydroxides are similar to oxides, being metals combined with the hydroxl ion, $(OH)^-$, rather than the O^{-2} ion. The hydroxides dissociate (break apart into ions) on heating. The $(OH)^-$ ions are driven off as water, as in the reaction $Mg(OH)_2 \rightarrow MgO + H_2O$.

The structure of an oxide or hydroxide is strongly dependent on the radii of the ions that form it.

ICE, H_2O (hydrogen oxide), is not normally thought of as a mineral, but it meets the definition of one. Unique in many ways, it is one of the few compounds that expands on freezing. This is fortunate, for otherwise ice would not float but would sink to the bottom, where it would tend not to melt. Under such conditions it is doubtful that life could have developed in the seas and ultimately given rise to life on land. Because it is solid below 0° C. at atmospheric pressures, it is abundant at the surface of the earth. As is well known, water has a high vapor pressure under normal atmospheric conditions and thus evaporates readily. Ice also has an appreciable vapor pressure and will change to the vapor by sublimation even well below 0° C.

Ice crystallizes in hexagonal system. As snow it forms complex stellate crystals that vary in pattern with changes in temperature. Snow compacts and recrystallizes to massive ice, which has interlocking irregular crystals. Snow crystals form readily on clay particles because of similar surface structures. Hardness is 1.5, specific gravity 0.92.

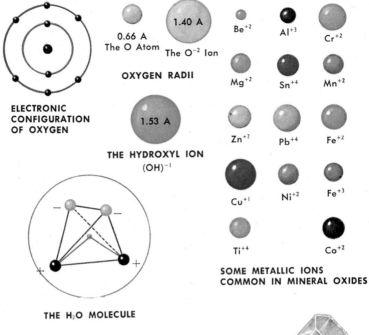

0.66 A
The O Atom

1.40 A
The O⁻² Ion

OXYGEN RADII

1.53 A

THE HYDROXYL ION
(OH)⁻¹

ELECTRONIC
CONFIGURATION
OF OXYGEN

Be^{+2}

Al^{+3}

Cr^{+2}

Mg^{+2}

Sn^{+4}

Mn^{+2}

Zn^{+2}

Pb^{+4}

Fe^{+2}

Cu^{+1}

Ni^{+2}

Fe^{+3}

Ti^{+4}

Co^{+2}

SOME METALLIC IONS
COMMON IN MINERAL OXIDES

THE H₂O MOLECULE

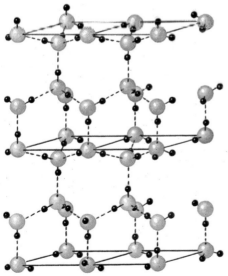

THE STRUCTURE OF ICE

Hollow Prism

Hexagonal Plate

Dendrite Snow
Crystals

PERICLASE GROUP includes all oxides that have bivalent (two charges) metallic ions and the halite structure (p. 120). In this structure, each metallic ion is adequately shielded by six oxygen ions around it. Only metallic ions of intermediate size assume this structure. Smaller ions are surrounded by four oxygen ions, larger ones by six. Oxides in which the bonding is more covalent have unique structures determined by the configuration of the electrons in the atom as well as the radius of the ion. Periclase-group minerals, like most oxides, are very reactive and so are found in unusual geologic environments. Of the five minerals in the group, only periclase is common. Two, CdO and CaO, are not described.

PERICLASE, MgO (magnesium oxide), like all oxides, is found in geologic environments that have either a deficiency of silica (otherwise silicates would form, p. 128) or high temperatures (at which silicates are unstable). It occurs as a metamorphic mineral in marbles formed by breakdown of dolomite and other magnesium minerals. It easily alters to brucite, $Mg(OH)_2$, under atmospheric conditions. Because of its high melting point, MgO is important as a refractory material. The large quantities of commercial MgO, however, are produced by heating magnesite, $MgCO_3$, or by treating $MgCl_2$, obtained as a brine from wells drilled into salt beds. The chemical similarly between Mg^{+2} and Fe^{+2}, noted in the silicates is evidenced by the fact that FeO and MgO form a complete solid solution series.

Periclase crystallizes in cubic system, forming octahedral or cubic-octahedral crystals, or irregular grains. It is brittle; transparent; glassy; colorless or, if impure, white, yellow, black, or green, with white streak; soluble in hydrochloric and nitric acids. It has perfect cubic cleavage. Hardness is 5.5, specific gravity 3.58. Sources: Italy, Sardinia, Spain, Sweden, Czechoslovakia, California, New Mexico.

BUNSENITE, NiO (nickel oxide), is commonly found in slags, but as a mineral has been reported only from a nickel-uranium vein at Johanngeorgenstadt, East Germany, formed as an oxidation product of nickel-cobalt arsenides. As a mineral it is insignificant, but its similarity to other periclase-group minerals is interesting and important. It forms octahedral crystals, sometimes in twins, in cubic system. It is a glassy dark green with brown-to-black streak. It can be dissolved with difficulty in acids. Hardness is 5.5, specific gravity 6.79. Can be synthesized by heating Ni metal in air.

MANGANOSITE, MnO (manganese oxide), has been found only in Sweden, with other manganese minerals and periclase in a high-temperature dolomite (marble), and at Franklin, N. J., with zinc minerals. The Franklin specimens are intergrown with zincite, ZnO. Manganosite crystallizes in cubic system, occurring as octahedral crystals, irregular grains, and masses. It has fair cubic cleavage. It is a transparent, glassy emerald green, tarnishing to black, with brown streak. It can be dissolved with difficulty in concentrated hydrochloric or nitric acid. Hardness 5.5, specific gravity 5.36.

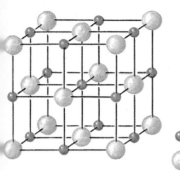

THE PERICLASE STRUCTURE

The structure of periclase, MgO, is the same as that of halite, NaCl (p. 121) because the cations in both cases can be adequately shielded by 6 anions.

Mg

O

PERICLASE CRYSTALS
(Idealized)

Massive Periclase
Langbanshyttan, Sweden

Crystal (Idealized)

Powder (Synthetic)

BUNSENITE

Manganosite
Franklin, N.J.

ZINCITE GROUP includes zincite, ZnO, and bromellite, BeO. Not abundant as minerals, these have a crystal structure assumed by many synthetic compounds. The metal ion, Zn^{+2} or Be^{+2}, is small enough to be adequately shielded by four oxygen ions, O^{-2}, around it. Each O^{-2}, in turn, is surrounded by four Zn^{+2} or Be^{+2}. The resulting hexagonal crystal shows a hemimorphic habit: its two ends have different faces, indicating a lack of a center of symmetry. Despite the identical structures of bromellite and zincite, bromellite is much harder because its smaller Be ion forms a more strongly covalent bond (p. 24) with oxygen.

ZINCITE, ZnO (zinc oxide), is a relatively rare mineral found with willemite and franklinite in calcite gangue in the zinc deposits at Franklin and Sterling, N. J. It is typically orange-yellow, though pure ZnO, prepared chemically, is white.

Zincite crystallizes in hexagonal system. It forms pyramidal crystals, commonly twinned base to base, and foliated or granular masses. It is brilliant in luster, has orange-yellow streak, is brittle, contains some iron and manganese, and is soluble in acids. Cleavage is prismatic, hardness 4, specific gravity 5.69. Zincite occurs in Poland, Italy, Spain, Australia (Tasmania), and New Jersey.

BROMELLITE, BeO (beryllium oxide), is unimportant as a mineral, the only significant beryllium ore being beryl. But synthetic bromellite has proved important as a refractory material for specialized purposes. Bromellite has been found at Lang- ban, Sweden, in a calcite vein of iron-rich rock. It occurs as minute crystals in hexagonal system and is transparent, white, and brittle. It dissolves in acids with difficulty. Cleavage is prismatic, hardness 9, specific gravity 3.04.

CUPRITE, Cu_2O (cuprous oxide), has a crystal structure different from zincite-group oxides. Each copper atom is surrounded by only two oxygen atoms. Ruby copper, as cuprite is often called, is a common mineral found as an oxidation product of copper sulfides in the upper zones of veins. It is usually associated with iron oxides, clays, malachite, azurite, and chalcocite. This and its color, crystal form, luster, and streak distinguish it from other minerals. It oxidizes to CuO in air.

Cuprite crystallizes in cubic system as cubes, octahedra, dodecahedra, and combinations of them; also as needles and fibrous and earthy masses. Brittle, it is red to black, with brilliant to earthy luster and red-brown streak. It gives white precipitate of cuprous chloride when dissolved in hydrochloric acid. Found in U.S.S.R., France, England, Australia, Congo, Chile, Bolivia, Mexico, Arizona, New Mexico, Nevada, California, Idaho, Colorado, Pennsylvania, Tennessee.

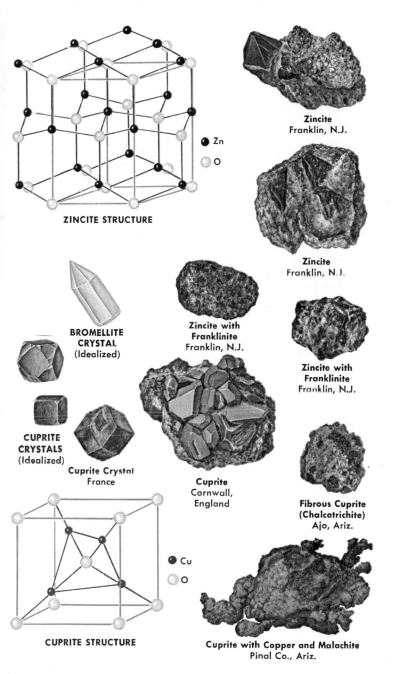

ZINCITE STRUCTURE

Zn
O

Zincite
Franklin, N.J.

Zincite
Franklin, N.J.

BROMELLITE CRYSTAL
(Idealized)

Zincite with Franklinite
Franklin, N.J.

Zincite with Franklinite
Franklin, N.J.

CUPRITE CRYSTALS
(Idealized)

Cuprite Crystal
France

Cuprite
Cornwall, England

Fibrous Cuprite (Chalcotrichite)
Ajo, Ariz.

CUPRITE STRUCTURE

Cu
O

Cuprite with Copper and Malachite
Pinal Co., Ariz.

133

CORUNDUM GROUP, commonly called the hematite group, includes the simple oxides Al_2O_3 (corundum) and Fe_2O_3 (hematite) and three mixed oxides of the general formula ABO_3. All have a rhombohedral (hexagonal) structure (p. 40), with each metallic ion surrounded by six oxygen ions. The metals may be trivalent, such as Al^{+3}, or may be mixed bivalent and tetravalent, such as the Fe^{+2} and Ti^{+4} in ilmenite. Much solid solution (p. 32) is possible among group members, as evidenced by ilmenite and pyrophanite. Members of the group are typically formed at high temperatures as disseminated grains in most igneous rocks or, in rarer cases, as large masses apparently segregated from the crystallizing magma at high temperatures. They are also common products of high-temperature metamorphism of silica-deficient rocks. Ilmenite, geikielite, and pyrophanite are far less common than corundum and hematite, but are interesting because they are closely related to the latter two in structure.

CORUNDUM, Al_2O_3 (aluminum oxide), is a common mineral important as an abrasive and as gemstones. It is found in metamorphosed bauxites and other aluminous rocks, in syenites and similar silica-poor igneous rocks, and in sediments as placers derived from such rocks. Corundum is associated with other oxides, with high-temperature accessory minerals such as zircon, and with aluminous silicates such as kyanite. Other forms of Al_2O_3 have been synthesized, but corundum is the most stable form and the only mineral form. Large gem-quality crystals of colored corundum are easily made by fusion of powdered Al_2O_3.

Corundum crystallizes in rhombohedral system, forming pyramids, prisms (often rounded into barrel shapes), and granular masses. It is translucent to transparent and is colorless, brown, red, blue, white, black, green, or gray. Its great hardness (9) distinguishes it, along with brilliant to glassy luster and high specific gravity (about 4). Common corundum occurs chiefly in Brazil, Malagasy Republic, South Africa, Ontario, Georgia, California; gems in Ceylon, Burma, Thailand; emery (a black variety) in Greece, Turkey, New York, Massachusetts.

CORUNDUM GEMSTONES are transparent crystals, free of flaws and deeply colored. Their great hardness, second only to diamond, makes them very durable. The relative rarity of good stones with the proper colors has increased their value. Pure corundum is colorless, but small amounts of metallic elements, present as impurities, impart various colors. The red variety, ruby, and the blue variety, sapphire, are among the most valuable. The yellow variety, called oriental "topaz," is not related to true topaz (sodium aluminum fluoride), nor is the purple oriental "amethyst" related to true amethyst (purple quartz). A striking starlike (asteriated) effect is produced in some stones by dispersion of light about the six-fold axis of the crystal. Such star rubies and star sapphires are among the most valued of gemstones.

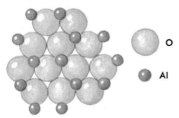

Hexagonal Arrangement of Atoms in Corundum

O
Al

CORUNDUM CRYSTALS
(Idealized)

Corundum in Feldspar
Renfrew, Ontario

Corundum
White Mts., Calif.

Corundum
Brazil

Ruby
Norway

Ruby in Zoisite
Africa

Blue Sapphire (x 10)

Star of India Sapphire

SYNTHETIC CORUNDUM

Aquamarine

Blue Sapphire

Ruby (x 10)

Sapphire
Ceylon

Pink Sapphire

Ruby

HEMATITE, Fe_2O_3 (ferric oxide), is the major ore of iron. Large deposits were mainly laid down as sediments, then altered by ground water and sometimes metamorphism. Three distinct types are recognized, each having the blood red to brownish red streak that is the chief distinguishing characteristic of hematite. Specular hematite occurs as brilliant black, thick or thin tabular crystals, commonly foliated. It is polished as a gemstone even though it is easily scratched or chipped. Red hematite, which may be very dark, is commonly found in columnar or radiating masses and fibrous clusters. Some kidney-shaped masses are called "kidney ore." Earthy, or ocherous, hematite is red or yellowish with a dull, earthy appearance, occasionally oolitic (in small round grains cemented together). It is commonly mixed with clays and sand.

Hematite crystallizes in rhombohedral system, sometimes forming twin crystals. It is brittle and opaque, contains some titanium, alters readily to limonite (p. 150), and is soluble in concentrated hydrochloric acid. Its hardness is 5-6, specific gravity 5.26. Vast deposits occur in Lake Superior area and in Appalachian Mts., New York to Alabama. Other notable sources: U.S.S.R., Switzerland, Italy, England, and Brazil.

ILMENITE, $FeTiO_3$ (iron titanate), is the major ore of titanium. Half the metal positions in its corundum-type structure (p. 134) are occupied by Ti^{+4} and the other half by Fe^{+2} (not Fe^{+3}, as in hematite). It occurs as veins or disseminated grains in basic igneous rocks (gabbros). Many sulfide veins and pegmatites also contain ilmenite. Placer deposits, notably black beach sands in Florida and India, are important sources.

Ilmenite crystallizes in rhombohedral system, forming rhombohedral, often platelike crystals (sometimes in twins) and compact masses, grains and sands. It is opaque, black (with black streak), metallic in luster, brittle, and magnetic. It enters into extensive solid solution with pyrophanite and is soluble in hot hydrochloric acid. Hardness is 5-6, specific gravity 4.79, varying with composition. Found in India, Australia, U.S.S.R., Norway, Sweden, Switzerland, France, Italy, England, Canada, Massachusetts, Rhode Island, Connecticut, New York, Pennsylvania, Kentucky, Florida, Idaho, Wyoming, and California.

GEIKIELITE, $MgTiO_3$ (magnesium titanate), is found in the gem gravels of Ceylon, but in few other localities. It occurs as opaque brownish black grains (in rhombohedral system) with brownish black streak and submetallic luster. It may contain much iron. Hardness is 5-6, specific gravity 3.97.

PYROPHANITE, $MnTiO_3$ (manganese titanate), is a rare mineral found as cavity fillings in veins. It crystallizes in rhombohedral system, is deep red, and has submetallic luster. Streak is brownish red or yellow with green tint; hardness 5-6; specific gravity 4.58. May contain much iron.

HEMATITE
CRYSTALS
(Idealized)

Specular Hematite
as a Gemstone

Earthy Hematite
Marampa, Sierra Leone

Specular Hematite
Humboldt, Mich.

Oolitic Hematite
Birmingham, Ala.

Red Streak

Earthy (Banded) Hematite
Mesabi Range, Minn.

Hematite, Kidney Ore
Westmoreland, Pa.

Ilmenite Crystal
Kragero, Norway

Massive Ilmenite
Mitchell Co., N.C.

Ilmenite Sand
N.Y.

ILMENITE-PYROPHANITE
CRYSTALS
(Idealized)

Ilmenite and Magnetite
Africa

SPINEL GROUP is perhaps the most complex group of oxides. A large number of natural species is recognized, and there is extensive solid solution (p. 32) among them. Many compounds with the spinel structure have been synthesized, including not only the mineral compounds but many others. Spinels are "mixed oxides." They contain two or more metals and have the general formula AB_2O_4, A and B designating the metals. Some metals are stable when surrounded by four oxygen ions (tetrahedral coordination) and others are stable when surrounded by six (octahedral coordination). Still others can occur in both tetrahedral and octahedral positions. The oxygen ions are in a cubic close-packed arrangement, with the metals located as shown on facing page.

SPINEL, $MgAl_2O_4$ (magnesium aluminum oxide), like all spinels, is a mineral formed at high temperature in igneous rocks (particularly silica-poor rocks) and in metamorphic rocks. It is not found in silica-rich rocks such as granites because cordierite, $Mg_2Al_4Si_5O_{18}$, is formed instead. Though pure $MgAl_2O_4$ is colorless, natural specimens may be red, yellow, blue, green, or many intermediate shades, depending on the presence of other metals. Clear, flawless stones are widely used as gems, and synthetic stones of gem quality are relatively inexpensive. Natural spinel may contain much iron, zinc, or chromium, and small amounts of other metals. Its hardness (7.5-8), glassy to dull luster, and octahedral crystals (when present) help identify it.

Spinel crystallizes in cubic system as octahedral crystals (commonly twinned), grains, and masses. It is transparent if pure, is brittle, and has white streak. It can be dissolved with difficulty in sulfuric acid. Specific gravity is 3.55, varying with composition. Found chiefly in U.S.S.R., Italy, Ceylon, Burma, **India,** Malagasy Republic, Canada, New York, New Jersey, Massachusetts, North Carolina, Alabama.

GAHNITE, $ZnAl_2O_4$ (zinc aluminum oxide), is relatively rare, but is interesting because it can form in silica-rich rocks such as granite pegmatites. (There is no stable zinc aluminosilicate that can form instead.) It occurs also in high-temperature metamorphic rocks, notably in the zinc deposits of Franklin, N. J. It forms masses and striated octahedral crystals in cubic system, is dark blue-green, with gray streak, and has glassy to dull luster. Hardness is 7.5-8, specific gravity 4.62. Occurs in W. Germany, Sweden, New Jersey, North Carolina, Massachusetts, and elsewhere.

HERCYNITE, $FeAl_2O_4$ (iron aluminum oxide), is quite rare. It is found as small grains in metamorphic rocks, associated with corundum, magnetite, garnets and various metamorphic minerals formed at high temperature. It is common in emery and has been found in diamond and cassiterite placers. It occurs as grains and masses in the cubic system, is black, and has glassy to dull luster and dark green streak. Hardness is 7.5-8, specific gravity 4.39. It occurs chiefly in W. Germany, Switzerland, India, Malagasy Republic, Australia (Tasmania), Brazil, New York, and Virginia.

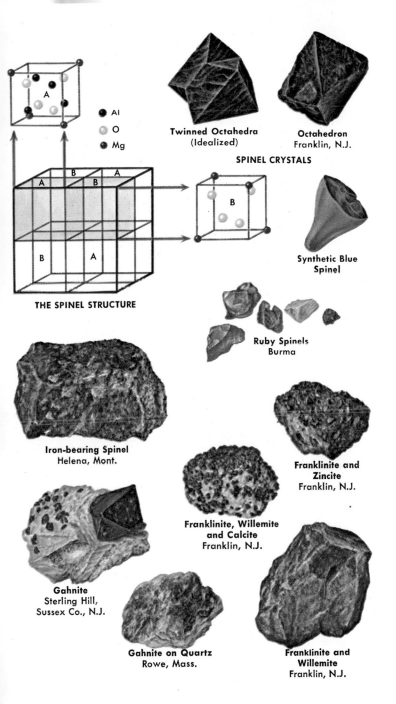

A

- Al
- O
- Mg

Twinned Octahedra
(Idealized)

Octahedron
Franklin, N.J.

SPINEL CRYSTALS

THE SPINEL STRUCTURE

Synthetic Blue Spinel

Ruby Spinels
Burma

Iron-bearing Spinel
Helena, Mont.

Franklinite and Zincite
Franklin, N.J.

Franklinite, Willemite and Calcite
Franklin, N.J.

Gahnite
Sterling Hill,
Sussex Co., N.J.

Gahnite on Quartz
Rowe, Mass.

Franklinite and Willemite
Franklin, N.J.

MAGNETITE, $FeFe_2O_4$ (ferrous and ferric iron oxide), contains both Fe^{+2} (ferrous) and Fe^{+3} (ferric) iron ions. Since the ferrous iron readily oxidizes to the ferric state under atmospheric conditions, specimens commonly are coated with hematite (p. 136) or, more often, limonite (p. 150). Magnetite is a valuable iron ore, large deposits of which were probably segregated from igneous magmas at high temperatures. Small grains occur in almost all igneous and metamorphic rocks. Magnetite is the most magnetic mineral. Some deposits, called lodestone, are permanently magnetized. Its magnetism, black color, hardness (5.5-6.5), and black streak are diagnostic.

Magnetite crystallizes in cubic system as octahedral and dodecahedral crystals, often twinned, and granular masses. It is brittle, opaque, and metallic to dull in luster; dissolves slowly in hydrochloric acid; has specific gravity of 5.20. Large deposits are found at Kiruna, Sweden, and in Adirondack region of New York. Also occurs in Lake Superior area of U.S. and Canada and in Norway, W. Germany, Italy, Switzerland, South Africa, India, Mexico, Oregon, New Jersey, Pennsylvania, North Carolina, Virginia, New Mexico, Utah, Colorado.

CHROMITE, $FeCr_2O_4$ (ferrous chromic oxide), is the only ore of chromium. It is the chromium analog of magnetite, and the two are similar in appearance. Chromite, however, is only weakly magnetic and has brown (rather than black) streak. Found with peridotite, a rock rich in olivine, it is probably one of the earliest minerals to crystallize from an igneous melt.

Chromite crystallizes in cubic system as octahedral crystals and granular masses. It is brittle, black, metallic, and opaque, with hardness of 5.5, specific gravity of 5.09. It is insoluble in acids. Only U.S.S.R., South Africa, Turkey, Philippines, Cuba, and Rhodesia have important deposits. Also found in New Caledonia, India, France, Yugoslavia, Bulgaria, Canada, Australia, California, Oregon, Maryland, Pennsylvania, Texas, Wyoming, North Carolina.

HAUSMANNITE, $MnMn_2O_4$ (manganese sesquioxide), occurs with magnetite and manganese minerals in veins and metamorphic rocks. It crystallizes in tetragonal system as pyramids and poorly developed prisms, often twinned, and granular masses. It is brittle, brownish black, submetallic in luster; has brown streak; is soluble in hot hydrochloric acid. Occurs in Bulgaria, Switzerland, England, Italy, Scotland, Sweden, India, and other areas.

OTHER SPINELS are less common. Galaxite, $MnAl_2O_4$, magnesiochromite, $MgCr_2O_4$, franklinite, $ZnFe_2O_4$, jacobsite, $MnFe_2O_4$, and trevorite, $NiFe_2O_4$, occur in small amounts and illustrate the wide variety of compositions possible. Extensive solid solution (p. 32) among these compounds further adds to the complexity of the spinel group. A large number of other compounds with the spinel structure have been synthesized from the component oxides.

MAGNETITE CRYSTALS
(Idealized)

Crystalline Magnetite
Mineville, N.Y.

**Massive Magnetite
with Red Hematite
Texas**

Lodestone
Magnet Cove, Ark.

Magnetite Sand
os Angeles Co., California

Magnetite Crystals
Austria

Magnetite
Espanola, Ontario

Magnetite in Jade
California

Emery
Chester, Mass.

Chromite
Lancaster Co., Pa.

Chromite
Transvaal,
South Africa

**HAUSMANNITE
CRYSTALS**
(Idealized)

Hausmannite
Germany

**Chromite (Black) with
Stichtite (Lilac) in
Serpentine (Green)**
Tasmania, Australia

141

RUTILE GROUP includes oxides with the formula MO_2, in which the metal (M) ion, with four positive charges, is adequately shielded by six oxygen ions. The result is the rutile structure. The ions of titanium, manganese, tin, and lead form oxides in this structure. But because of chemical differences in the ions, their oxides exhibit little solid solution and little similarity in origin. Rutile, anatase, and brookite are three forms (polymorphs) of titanium dioxide that have the same composition but different arrangements of the rutile structure. They are minor ores of titanium, the compound TiO_2 being more important as the white pigment in paints and porcelain enamels.

RUTILE, TiO_2 (titanium dioxide), is a common, widespread mineral, but occurs in small amounts. It is an accessory mineral in granites, forming early in the crystallization sequence; in metamorphic rocks of all types; and especially in veins, where it is intimately associated with quartz. Quartz crystals laced with needles of rutile (rutilated) are common in veins of igneous rocks. Rutile is very resistant to weathering and so is a common constituent of sediments. Hardness and high index of refraction make it an ideal gemstone. Synthetic rutile crystals are clear. Rutile is the only form of TiO stable at all temperatures, but anatase is commonly the first form to crystallize from a porcelain enamel. It then changes to rutile if held long enough at a high temperature.

Rutile crystallizes in tetragonal system, forming prisms that may be needlelike, may have pyramid ends, and may be joined as elbow twins. Granular masses are also common. Rutile is red to brown or black, with streak of brown, yellow, gray, or greenish black, depending on composition (contains much iron, some tantalum and nickel). It is brilliant in luster, transparent to opaque, and brittle. Cleavage is prismatic, hardness 6-6.5, specific gravity 4.26. Occurs in Norway, Sweden, Australia, U.S.S.R., Italy, France, Malaysia, Malagasy Republic, Brazil, Arkansas, Virginia, North Carolina, South Dakota, California, and other areas.

ANATASE, TiO_2 (titanium dioxide), occurs typically in veins in metamorphic rocks; also in igneous and sedimentary rocks. It commonly is found with rutile. It crystallizes in tetragonal system as pyramidal or tabular crystals, rarely twins. It exists in many colors, is brilliant in luster, is transparent to opaque, and has colorless or pale streak. It changes to rutile at high temperatures and is found in about the same places. Hardness 5.5-6, specific gravity 4.04.

BROOKITE, TiO_2 (titanium dioxide), is found with anatase and rutile in similar environments and localities. Like anatase it converts to rutile at high temperatures, but has not yet been conclusively proved to be a true TiO_2 polymorph. It crystallizes in orthorhombic system as tabular or prismatic crystals, rarely twins. It has many colors, is brilliant in luster, has colorless or pale streak, and is transparent only in small pieces. Cleavage is indistinct, hardness 5.5-6, specific gravity 4.12.

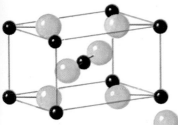

RUTILE STRUCTURE
(unit cell)

 O

Ti

RUTILE CRYSTALS
(Idealized)

Rutile Crystal
Roseland, Va.

Rutile Needles
Lincolnton, N.C.

**Rutile,
var. Nigrine**
Magnet Cove, Ark.

**Rutile in
Pegmatite**
Nelson Co., Va.

Rutile Crystals
Lincoln Co., Ga.

ANATASE CRYSTAL
(Idealized)

Anatase
Binnental,
Switzerland

Brookite Crystal
Switzerland

Brookite
Magnet Cove, Ark.

143

PYROLUSITE, MnO_2 (manganese dioxide), is a secondary material formed when water leaches manganese from igneous rock and redeposits it as concentrations of manganese dioxide. As a result it occurs more often as coatings on other minerals than as large crystals. Nodules of pyrolusite are found in many areas of the ocean floor and may prove valuable in the future. Manganese is very important in the manufacture of steel and in many other industrial processes.

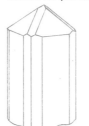

Pyrolusite crystallizes in tetragonal system, rarely as prismatic crystals, commonly as powdery, granular, crystalline, or fibrous masses, or dendritic crusts. It is bluish gray to black, with black streak; metallic to earthy in luster; opaque; and brittle. It dissolves in hydrochloric acid, with acrid chlorine gas given off. Cleavage is prismatic, hardness 6-6.5 (less if massive or powdery), specific gravity 5.24. Occurs in W. Germany, U.S.S.R., India, Brazil, Cuba, Virginia, Tennessee, Georgia, Arkansas, Minnesota, and many other places.

CASSITERITE, SnO_2 (tin dioxide), is found in small amounts in many areas, normally associated with silica-rich rocks. It is a minor constituent of granite, particularly pegmatite, and commonly occurs in veins associated with quartz near granite bodies. It is also found as an oxidation product in tin-bearing sulfide veins. Because it is very heavy, it is mined primarily from placer deposits, notably on the Malay Peninsula. It is the major ore of tin, which is valuable as a coating for steel and as a constituent of soft alloys such as solder.

Cassiterite crystallizes in tetragonal system as short prismatic or pyramidal crystals, often twinned in elbow shapes, and as masses, grains, crusts, and radiating fibers. It is opaque to transparent black, yellow, red, or white, with white, gray, or brown streak and brilliant luster. Contains much iron, some tantalum, niobium. Major producers are Malaysia, Indonesia, Bolivia, Congo, and Nigeria; occurs also in East Germany, England, Mexico, Virginia, South Dakota, New Mexico, California, South Carolina, Maine, New Hampshire, Texas, etc.

PLATTNERITE, PbO_2 (lead dioxide), is a relatively rare mineral found as an oxidation product of lead-bearing vein deposits. In contrast with the other oxides, it is brownish black to jet black. Crystallizing in the tetragonal system, it commonly occurs as nodules or powdery masses associated with limonite; also as prismatic crystals, sometimes twinned, and fibrous masses. It is brittle, has brilliant metallic luster that dulls on exposure, is opaque, and has brown streak. It contains some zinc, copper, iron; dissolves in sulfuric acid with precipitation of lead sulfate (white), and in hydrochloric acid with evolution of chlorine gas (greenish yellow). Hardness is 5.5, specific gravity 9.63. Occurs in Scotland, South-West Africa, Mexico, Idaho, and South Dakota.

PYROLUSITE CRYSTALS
(Idealized)

Pyrolusite
Nova Scotia

Pyrolusite
Lake Valley, N. Mex.

**Manganese Ore
with Pyrolusite**
Amapa, Brazil

**CASSITERITE
CRYSTALS**
(Idealized)

**Cassiterite (Black) in
Altered Marble**
Keystone, S. Dak.

Cassiterite
Taylor Creek, N. Mex.

**PLATTNERITE
CRYSTAL**
(Pseudomorph
after
Pyromorphite)
(Idealized)

Plattnerite
Mina Ojuela, Mapimí, Mexico

**Cassiterite
Grains**
Hill City, S. Dak.

145

URANINITE GROUP includes uraninite and thorianite, oxides of uranium and thorium. They have a cubic structure of the fluorite type (p. 124), with each uranium or thorium ion surrounded by eight oxygen ions. Since complete solid solution is possible between uraninite and thorianite, they can form a whole range of minerals intermediate in composition. Uraninite is valuable as a source of uranium for atomic fission. Although uranium is found in most igneous rocks, it occurs there in such minute and widely dispersed amounts that it is of little economic value.

URANINITE, UO_2 (uranium dioxide), is the major ore of uranium. It occurs as crystals in granite and syenite pegmatites, associated with rare earth minerals, and as masses in hydrothermal sulfide veins. The isotopes of uranium are radioactive, decaying at a known rate to lead and helium, which are always found in uraninite. The lead-uranium or helium-uranium ratio, which increases with time, is used to determine the mineral's age.

Uraninite crystallizes in cubic system as octahedra, cubes, or combinations, and as often botryoidal masses (called pitchblende). It is brittle; black to brownish or purplish black, often altered to hydrates of various colors; submetallic, greasy, or dull in luster; and opaque; with black, brownish, gray, or olive streak. It is generally oxidized, with actual composition between UO_2 and U_3O_8. It is soluble in sulfuric, nitric, and hydrofluoric acids. Hardness is 5-6, specific gravity 8-10.88 (variable). Occurs notably in West Germany, England, South Africa, Congo, Canada, New Hampshire, Connecticut, North Carolina.

THORIANITE, ThO_2 (thorium dioxide), is the thorium analog of uraninite. It is the major ore of thorium. Largest deposits are placers, found in Ceylon, Malagasy Republic, and U.S.S.R. It also occurs in serpentine at margins of a pegmatite at Easton, Pa. Most specimens contain much uranium, also cesium and lanthanum. Thorianite crystallizes in cubic system as cubic crystals, commonly rounded. It is opaque black, gray, or brownish, with gray or greenish streak, is submetallic or horny in luster, and is brittle. It alters easily to gummite (p. 150) and is soluble in sulfuric and nitric acids. Hardness is 6.5, specific gravity 9.87 (variable). Thorianite, unlike uraninite, does not oxidize to Th_3O_8. Thorium is the luminous material in gas mantles, used in camping lanterns.

CHRYSOBERYL, $BeAl_2O_4$ (beryllium aluminum oxide), is not related to the uraninite group, but is a mineralogical curiosity. Though its formula suggests the spinel ($MgAl_2O_4$) structure, it assumes the structure of olivine (Mg_2SiO_4). It occurs beryllium-rich pegmatites and in mica schists and marbles. Gem varieties include cat's-eye, which exhibits chatoyancy, and alexandrite, which is emerald green but is red by transmitted light and in artificial light. Chryoberyl crystallizes in orthorhombic system, forming crystals that often are twinned. It is transparent green, yellow, or brown; streak colorless. Hardness is 8.5, specific gravity 3.69. Occurs chiefly in U.S.S.R., Malagasy Republic, Ceylon, Brazil, Maine, Connecticut, New York, and Colorado.

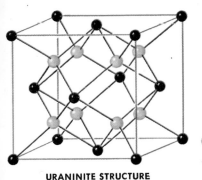

URANINITE STRUCTURE
Identical with Fluorite (p. 125)

O
U

URANINITE CRYSTALS
(Idealized)

**Uraninite
with Gummite**
Grafton Center, N.H.

Purple Uraninite
Bancroft, Ontario

Massive Uraninite
Ike Mine, La Sal Mts.,
Utah

Thorian Uraninite
Haliburton, Ontario

**Uraninite with Gummite
and Uranophane**
Mitchell Co., N.C.

Thorianite Crystals
Balangoda, Ceylon

Chrysoberyl
Haddam, Conn.

**Faceted
Chrysoberyl**

Chrysoberyl
Lake Alaotra,
Malagasy Republic

DIASPORE GROUP includes those oxyhydroxides of trivalent metals, Al^{+3}, Fe^{+3}, and Mn^{+3}, that have each metal ion surrounded by six negative ions—three oxygen, O^{-2}, and three hydroxide, $(OH)^-$. Diaspore-group minerals are only three of many hydroxides, some noncrystalline, that are important sources of these metals. They are associated with other hydrous minerals.

DIASPORE, $AlO(OH)$ (aluminum oxyhydroxide), is found as a secondary product formed by alteration of corundum and other aluminous minerals by hydrothermal solutions in igneous and metamorphic rocks or by chemical weathering. It is a major constituent of bauxites. When heated, it loses water and converts to corundum (p. 134).

Diaspore crystallizes in orthorhombic system as crystals of various forms, as foliated masses, and as disseminated grains. It is transparent white, gray, yellowish, or greenish, with white streak; greasy in luster, but pearly along platy cleavages; and brittle. It crackles when heated. Hardness is 6.5-7, specific gravity 3.37. Notable sources: Hungary, France, U.S.S.R., Switzerland, Massachusetts, Pennsylvania, Colorado, California, Arkansas, Missouri, and North Carolina.

GOETHITE, $FeO(OH)$ (iron oxyhydroxide), is derived by weathering from iron-bearing minerals. It crystallizes in orthorhombic system as tablets, scales, needles, radial and concentric aggregates, and earthy or botryoidal masses. It is opaque brown, blackish, or yellowish, with distinctive yellow streak; brilliant metallic to dull in luster; and brittle. Hardness 5-5.5, specific gravity 3.3-3.5. England, Cuba, Michigan, Minnesota, Colorado, Alabama, Georgia, Virginia, Tennessee.

MANGANITE, $MnO(OH)$ (manganese oxyhydroxide), is a minor ore of manganese, found in low-temperature veins and in deposits formed by weathering and groundwater alterations of other minerals. It crystallizes in monoclinic system as striated prisms, sometimes in bundles, often in elbow and penetration forms. It is brittle and gray or black, with reddish brown or black streak and submetallic luster. Hardness is 4, specific gravity 4.38. Occurs chiefly in West Germany, England, Nova Scotia, and Michigan.

BAUXITE, the principal ore of aluminum, is a mixture of diaspore, gibbsite, boehmite, (a cubic modification of diaspore), and other materials. Noncrystalline colloidal precipitates, clay, limonite, and partly weathered silicates may be present. The nature of bauxite and its purity vary widely. Commercial deposits, necessarily massive may be varicolored, but are generally gray or white with reddish-brown iron staining. Bauxite is commonly the result of prolonged weathering of aluminous rocks, notably syenites, in tropical and subtropical climates. Under these conditions the silica component of the rock dissolves and leaves the aluminum hydroxides as a residue. Under acid conditions of the temperate regions, aluminum is leached by weathering and the silica is left. Arkansas is the major U.S. producer, with smaller deposits in Georgia, Alabama, and Mississippi. France, Indonesia, U.S.S.R., Hungary, Guyana, and Venezuela also are major sources. It is quite commonly used as a source of Al_2O_3 for ceramics.

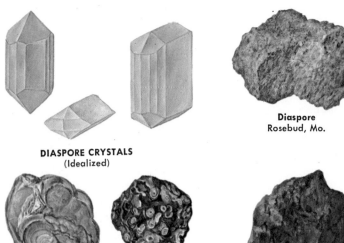

DIASPORE CRYSTALS
(Idealized)

Diaspore
Rosebud, Mo.

Gibbsite
Minas Gerais, Brazil

Earthy Goethite
Gypsum, Colo.

Manganite
Michigan

**Radiating
Goethite**
Biwabik, Minn.

**Botryoidal
Goethite,
Mexico**

BAUXITES, CONTAINING DIASPORE AND GIBBSITE

Bauxite
Demerara,
Guyana

Bauxite
Bauxite, Ark.

Bauxite
Little Rock, Ark.

BRUCITE GROUP includes the hydroxides of the divalent metals magnesium, Mg^{+2}, and manganese, Mn^{+2}. Structure is layer type with six hydroxyl ions, $(OH)^-$, surrounding each metal ion. Charges on metal ions are satisfied within the layers, which are held together by hydrogen bonds (p. 30).

BRUCITE, $Mg(OH)_2$ (magnesium hydroxide), is an alteration product of periclase, MgO, in magnesium-rich marbles and is a vein mineral formed at low temperature in serpentine or dolomite metamorphic rocks. Layers are flexible and easily separated.

Brucite crystallizes in hexagonal system as plates, often foliated, and fibrous masses. Its color varies from green (iron-rich varieties) to yellow, reddish, or brown (manganese-rich). It is transparent; pearly or waxy in luster; white soluble in acids. Up to 20% Mn, some Fe and Zn. Hardness is 2.5, specific gravity 2.40. Occurs in Austria, Italy, Scotland, Sweden, U.S.S.R., Canada, New York, Nevada, and California.

PYROCHROITE, $Mn(OH)_2$ (manganese hydroxide), is closely analogous to brucite, the major difference being the ease with which Mn^{+2} oxidizes to Mn^{+3}. Pyrochroite therefore weathers readily. It is found with calcite and dolomite as well as other manganese minerals in low-temperature hydrothermal veins. It crystallizes as flexible tabular crystals and masses and veinlets. It is colorless, pale green, or blue, altering to brown or black; has pearly luster; and is opaque. Much magnesium may substitute for Mn. Hardness is 2.5, specific gravity 3.25. Occurs in Sweden, Switzerland, Yugoslavia, New Jersey, and California.

RELATED MINERALS

LIMONITE, like bauxite, is not a mineral but a mixture of materials. Noncrystalline iron hydroxide may properly be called limonite. Limonite occurs as massive, crusty, stalactitic cavity fillins and as varnishlike coating on rocks. An important iron ore where abundant, it invariably occurs with hematite and goethite. It is formed by alteration of iron oxides, sulfides, and silicates. It is amorphous, is glassy to dull in luster, has yellowish brown streak. Hardness 1-5.5, specific gravity 2.7-4.3.

PSILOMELANE, $BaMnMn_8O_{16}(OH)_4$, is a complex but common manganese mineral. It is a weathering product of carbonates and silicates, found in swamp clays, Mn-rich veins, limestones, and metamorphic rocks. It crystallizes only as massive, fine-grained crusts, stalactites, and cavity fillings. It is opaque black, with brown to black streak; nearly metallic to earthy, brittle. Hardness 5-6, specific gravity 4.42. Chief sources: France, Belgium, Scotland, Sweden, India, Virginia, Arizona.

GUMMITE is a mixture of materials, largely amorphous, containing uranium, lead, and thorium in large amounts. It represents the weathering products of uranium oxide. It occurs as masses or crusts and as pseudomorphs after original minerals. It is yellow-orange, orange, brown, or black; greasy, waxy, glassy, or dull in luster. Hardness is 2.5-5, specific gravity 3.9-6.4. Occurs in Norway, Congo, South Africa, Canada, North Carolina, Pennsylvania, Connecticut, Maine.

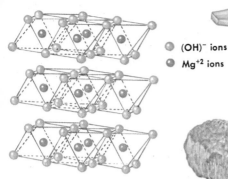

(OH)⁻ ions

Mg⁺² ions

BRUCITE CRYSTALS
(Idealized)

BRUCITE STRUCTURE

Brucite
Wakefield,
Quebec

Brucite
Nevada

**Limonite
Powder (Ochre)**
Cartersville, Ga.

Pyrochroite
Langban, Sweden

Limonite
Newport, N.Y.

Limonite
Wyoming

Limonite
Pelican Point,
Utah

**Limonite Alteration
of Pyrite**
Mexico

Psilomelane
Mexico

Psilomelane
Dana Co., N. Mex.

Gummite
Mitchell Co., N.C.

BORATES

Borates are a complex class of about 45 minerals. All contain boron and oxygen in chemical combination with metals. About four-fifths of them also contain water, and some contain hydroxyl ions, (OH), or halogen ions such as chlorine. Others are compound borate-phosphates, borate-sulfates, or borate-arsenates. The basic building block of most borates is a boron ion surrounded by three oxygens ions in the same plane, (BO_3). This boron triangle, like the silica tetrahedron (p. 156), can form infinite rings and chains. In some borates, such as borax, the boron is surrounded by three oxygen ions and one hydroxyl ion. Other possible combinations are shown on the facing page. Only five of the more common borates are described here. Boron fibers are important in some newly developed materials.

BORAX, $Na_2B_4O_5(OH)_4 \cdot 8H_2O$ (hydrous sodium borate), is the best known and most widespread borate. It occurs in large deposits in the dry beds of salt lakes in arid regions with other borates, halite, and gypsum. At atmospheric temperatures, clear crystals lose water and turn white (effloresce).

Borax crystallizes in monoclinic system, with two-directional cleavage, as prismatic crystals, crusts, and porous masses. It is colorless, white, or tinted, with white streak; glassy or resinous in luster; translucent to opaque; and brittle. It dissolves in water, producing a sweet alkaline taste. Hardness is 2-2.5, specific gravity 1.70. It occurs in India, Tibet, U.S.S.R., Iraq, California, Nevada, and New Mexico.

COLEMANITE, $Ca_2B_6O_{11} \cdot 5H_2O$ (hydrous calcium borate), is found with borax notably in California playa lakes. It crystallizes in monoclinic system as short prisms and granular masses. It is white or colorless, brilliant in luster, tasteless, and soluble in hot hydrochloric acid (white flakes appear on cooling). It is insoluble in water. Hardness is 4.5, specific gravity 2.42.

ULEXITE, $NaCaB_5O_9 \cdot 8H_2O$ (hydrous sodium calcium borate), commonly crystallizes in triclinic system as aggregates of radiating needlelike crystals that form rounded, white, silky masses called "cotton balls." It is associated with borax, notably in Chile, Argentina, Nevada, and California. Tasteless, it decomposes in hot water. Hardness is 2.5 or less, specific gravity 2.

BORACITE, $Mg_3B_7O_{13}Cl$ (magnesium borate), is found notably in salt domes in Louisiana and in potash deposits in East Germany, France, and England. It occurs as combinations of crystal forms and as fibrous or granular masses. It is white or various colors, glassy, and slowly soluble in hydrochloric acid. Hardness is 7-7.5, specific gravity 2.97.

SUSSEXITE, $(Mn, Mg)BO_2(OH)$ (manganese magnesium borate), occurs as fibrous veinlets or masses associated with manganese and zinc minerals. Crystal system is probably orthorhombic. Mineral is brittle, is white or buff, and dissolves slowly in acids. Occurs in Hungary, U.S.S.R., Korea, Sweden, New Jersey, Michigan, California, and Nevada.

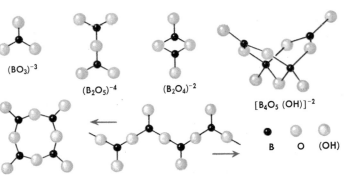

$(BO_3)^{-3}$

$(B_2O_5)^{-4}$

$(B_2O_4)^{-2}$

$[B_4O_5(OH)]^{-2}$

$(B_4O_8)^{-4}$

THE INFINITE CHAIN, $(BO_2)_n^{-n}$

● B ○ O ○ (OH)

BORATE IONS

Borax
Borax Lake, Calif.

Borax
San Bernardino, Calif.

Borax
Esmeralda Co., Nev.

Colemanite
Kern Co., Calif.

Ulexite
Kern Co., Calif.

Crystal

Boracite
Hannover, West Germany

Sussexite
Franklin, N.J.

153

SILICATES

Silicates are combinations of silicon and oxygen with one or more metals. Since silicon and oxygen together make up about 75 per cent of the earth's crust, silicates are by far the most abundant minerals. If quartz (silicon dioxide) is included with the silicates rather than the oxides, as justified by its characteristics, then about 93 per cent of the crust is composed of silicates.

Silicates are relatively difficult to identify because of their number and complexity. Though their characteristics vary, most silicates are transparent to translucent, are glassy in luster, are relatively hard, and are insoluble, or nearly so, in acids.

CHEMISTRY of the silicates depends largely on the silicon and oxygen atoms. The silicon atom has four electrons in its outer shell. It most readily obtains a stable outer shell (p. 20) in the presence of oxygen by losing the four outer electrons, thereby becoming the silicon ion, Si^{+4}. The oxygen atom has six electrons in its outer shell and most easily attains a stable outer shell by gaining two electrons, thus becoming the oxygen ion, O^{-2}. The positive Si^{+4} and the negative O^{-2} attract each other. The bond between them is a strong one, having appreciable covalent character, that affects the characteristics of silicates. The sizes of the ions are important in determining crystal structure and the temperature at which crystallization occurs.

BASIC BUILDING BLOCK of silicates is the silica tetrahedron, $(SiO_4)^{-4}$. It consists of the small silicon ion, Si^{+4}, surrounded by four oxygen ions, O^{-2}, which adequately shield it. The centers of the O^{-2} are the corners of a regular tetrahedron. The excess negative charge on the whole unit is four. The four positive charges of the Si^{+4} are evenly divided among the four O^{-2}. Thus each O^{-2} is half saturated. The remaining negative charge on each O^{-2} can be satisfied in various ways. If the O^{-2} is shared by two tetrahedra, for example, its charge is totally satisfied (it is saturated).

Unshared O^{-2} depend on metal ions other than silicon for saturation. These ions, being larger, are inadequately shielded by four O^{-2}. The iron ion, Fe^{+2}, and magnesium ion, Mg^{+2}, each must be surrounded by six O^{-2} whose centers form a regular octahedron. Since the ion's two positive charges are evenly divided among six O^{-2}, each O^{-2} receives a third of a charge. Thus an O^{-2} of a silica tetrahedron, with one charge satisfied by the Si^{+4}, needs three Fe^{+2} or Mg^{+2} around it to be completely saturated. The still larger sodium, calcium, and potassium ions must be surrounded by more than six O^{-2}.

Silicate minerals consist of ions, of finite size—independent tetrahedral, double tetrahedral, or ring silicate—or of infinite ions, similar to polymers—chain, sheet, and framework—packed into three dimensional arrays, held together by metal ions that saturate the unshared oxygens.

**Several Views of the
Silica Tetrahedron**

**Several Views of
the Octahedron**

**Two Tetrahedra Put
Together in Orientation
as Shown to Provide an
Octahedral Position**

**Single Tetrahedron—Oxygen
Ions One-half Saturated**

**Double Tetrahedral Unit—Only
the Shared Oxygen is Saturated**

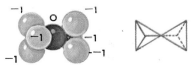

A Portion of a Silica Chain

Each tetrahedron shares two oxygens with
adjacent tetrahedra. The shared oxygens
are saturated. The others must be saturated
by metal ions between the chains.

155

CLASSIFICATION of silicates is based on the number of shared oxygen ions per silica tetrahedron. A tetrahedron may share from zero to all four of its oxygen ions with adjacent tetrahedra. This method of classification is convenient because the characteristics and behavior of silicates are largely dependent on the nature of the tetrahedral arrangements. In turn, the characteristics and behavior of silicates provide many clues to their internal structure. Silicates are divided into seven groups:

Independent tetrahedral silicates (p. 158) consist of silica tetrahedra that share no oxygen ions. The remaining charge on each oxygen ion is satisfied by metal ions between the tetrahedra. These ions hold the structure together. The tetrahedra, or $(SiO_4)^{-4}$ groups, are similar to complex negative ions.

Double tetrahedral silicates (p. 168) form when each tetrahedron shares one O^{-2} with another tetrahedron, resulting in $(Si_2O_7)^{-6}$ groups. These act as complex negative ions and are held together in crystals by positive ions that occupy positions between them and balance their charges.

Ring silicates (p. 170) consist of silica tetrahedra joined in rings, with each tetrahedron sharing two O^{-2} with adjacent tetrahedra. The rings are complex groups of $(SiO_3)_3^{-6}$ or $(SiO_3)_6^{-12}$. Charges on these may be balanced by other metal ions that hold the rings together in a crystal structure.

Single-chain silicates (p. 174), like ring silicates, are formed by each tetrahedron sharing two O^{-2}. In effect, a giant negative ion is created with an indefinite number of tetrahedra, each carrying two negative charges. The chains are aligned and held together by metal ions other than silicon.

Double-chain silicates (p. 182) are formed when half the tetrahedra share two O^{-2} and the other half share three O^{-2}. The resulting giant negative ion has an indefinite number of $(Si_4O_{11})^{-6}$ units. Chains are aligned and held together by metal ions.

Sheet silicates (p. 186) are formed when each tetrahedron shares three O^{-2} with other tetrahedra. The resulting giant negative ion extends indefinitely in two dimensions. The sheets consist of $(Si_2O_5)^{-2}$ units held together in stacks by metal ions. The perfect cleavage of mica is a direct result of this structure.

Framework silicates (p. 204) have tetrahedra that share all four O^{-2} with adjacent tetrahedra. The result is a framework extending indefinitely in three dimensions. If Si^{+4} ions occupy all tetrahedral positions, no other metal ions are necessary to saturate the O^{-2} ions, and quartz, SiO_2, forms. But if aluminum ions, Al^{+3}, occupy some tetrahedral positions, other ions—notably potassium, K^{+1}; sodium, Na^{+1}; and calcium, Ca^{+2}—can enter the framework between tetrahedra, as in the feldspars, $K (AlSi_3) O_8$, $Na (AlSi_3) O_8$, and $Ca (Al_2Si_2) O_8$.

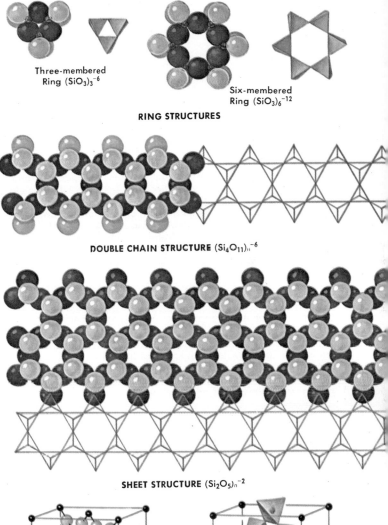

Three-membered
Ring $(SiO_3)_3^{-6}$

Six-membered
Ring $(SiO_3)_6^{-12}$

RING STRUCTURES

DOUBLE CHAIN STRUCTURE $(Si_4O_{11})_n^{-6}$

SHEET STRUCTURE $(Si_2O_5)_n^{-2}$

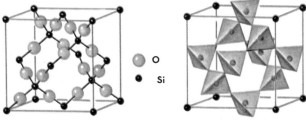

O   O

•   Si

FRAMEWORK STRUCTURE (SiO_2)

CRISTOBALITE

INDEPENDENT TETRAHEDRAL SILICATES

These silicates, also called orthosilicates because they were once assumed to be salts of orthosilicic acid, are made up of silica tetrahedra that share no oxygen ions. The negative tetrahedra are balanced by positive ions positioned among them in such a way that each positive ion is surrounded by the number of oxygen ions needed to shield it. The positive ions hold the resulting crystal together and determine its symmetry. In general, these silicates crystallize at high temperatures in metamorphic and igneous rocks.

ZIRCON GROUP has zirconium or thorium ions packed among silica tetrahedra in such a way that each Zr^{+4} or Th^{+4} ion has eight oxygen ions around it. Other M^{+4} ions (M = metal) are stable only when surrounded by fewer than eight oxygen ions, so only zircon and thorite are known to assume this structure, which is tetragonal, with the unequal axis slightly shorter than the two equal axes (p. 40).

ZIRCON, $ZrSiO_4$ (zirconium silicate), is found in nearly all igneous rocks and in some metamorphic rocks as small crystals, usually in the form of prism-pyramid combinations. These crystals, apparently formed at high temperatures, are widely and evenly distributed and generally make up less than 1% of the rock. Zircon also occurs as rounded grains in sandstones. Radioactive disintegration of trace elements in many zircons has destroyed the crystal structure (metamict mineral) and in some cases has produced colored haloes (pleochroism) in the surrounding mineral. Because of their high index of refraction and hardness, zircons are widely used as gemstones.

Zircon crystallizes in tetragonal system. It is transparent to opaque; colorless, yellow, gray, brown, blue, or green (no streak); brilliant in luster; brittle; and very resistant to weathering and chemical attack. Fracture is conchoidal, hardness **7.5**, specific gravity **4.2-4.9**. Occurrence is worldwide. Because of its hardness, brilliance, and high index of refraction, zircon has been used widely as a gemstone, but has not been highly valued because of its reputation as an imitation diamond.

THORITE, $ThSiO_4$ (thorium silicate), is quite rare. This thorium analog of zircon has been found, associated with other radioactive ores, in Norway, Sweden, Canada, and Malagasy Republic. It often contains uranium. Commonly hydrated, it weathers more readily than zircon. It crystallizes in tetragonal system, with good prismatic cleavage, as large or small zircon-like crystals and as grains. Its radioactivity often destroys crystal structure, making metamict varieties very common. It is black, brown, or orange; is brittle; and has hardness of 4.5 and specific gravity or 4.5-5.4, depending on hydration and composition.

CLASSIFICATION OF COMMON
INDEPENDENT TETRAHEDRAL SILICATES

ZIRCON GROUP
Zircon $ZrSiO_4$
Thorite $ThSiO_4$

EPIDOTE GROUP
Epidote $Ca_2(Fe,Al)_2O(OH)(Si_2O_7)(SiO_4)$
Zoisite
Clinozoisite $\Big\}$ $Ca_2Al(Al_3)O(OH)(Si_2O_7)(SiO_4)$
Piedmontite $Ca_2(Mn,Fe,Al)_3O(OH)(Si_2O_7)(SiO_4)$

GARNET GROUP
Almandite $Fe_3Al_2(SiO_4)_3$
Andradite $Ca_3Fe_2(SiO_4)_3$
Grossularite $Ca_3Al_2(SiO_4)_3$
Pyrope $Mg_3Al_2(SiO_4)_3$
Spessartite $Mn_3Al_2(SiO_4)_3$
Uvarovite $Ca_3Cr_2(SiO_4)_3$

THE OLIVINE GROUP
Forsterite Mg_2SiO_4
Fayalite Fe_2SiO_4
Tephroite Mn_2SiO_4
Monticellite $CaMgSiO_4$

SUBSATURATES

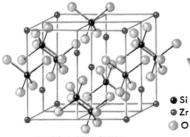

ZIRCON STRUCTURE
Showing Atoms

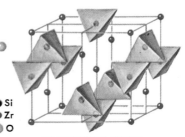

● Si
◉ Zr
○ O

TETRAHEDRAL MODEL
of Zircon

Zircon Crystals
Thailand

ZIRCON CRYSTALS
(Idealized)

Zircon Crystal
Essex Co., N.Y.

Zircon
Renfrew, Ontario

Thorite
Cardiff, Ontario

Thorite
Malagasy Republic

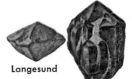

Langesund

Arendal

Thorite Crystals
Norway

159

EPIDOTE GROUP includes minerals with a complex structure containing both independent silica tetrahedra, $(SiO_4)^{-4}$, and double tetrahedra, $(Si_2O_7)^{-6}$. These are held together by the ions of aluminum, Al^{+3}, iron, Fe^{+3}, manganese, Mn^{+3}, and calcium, Ca^{+2}; water is present as hydroxyl ions, $(OH)^{-1}$. Members of the group are difficult to distinguish. They occur in generally small amounts in low- to medium-grade metamorphic rocks formed at high pressures and in igneous rocks as late-stage precipitates from magmatic fluids.

EPIDOTE, $Ca_2Fe(Al_2O)(OH)(Si_2O_7)(SiO_4)$ (hydrous calcium iron-aluminum silicate), is the iron-rich end member of a complete solid-solution series (p. 32) with clinozoisite, the aluminum-rich end member. It is common in calcium-rich metamorphic rocks and in sedimentary rocks near contacts with igneous bodies. Its typical pistachio green color and its cleavage (perfect in one direction, imperfect in another) help identify it.

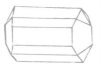

Epidote crystallizes in monoclinic system, forming prismatic crystals and fibrous or granular masses. Besides pistachio green, it may be emerald green, black, red, or yellow (with no streak); it is glassy, transparent to opaque, and brittle. Hardness is 6-7, specific gravity 3.2-3.5. Very widespread in metamorphic areas throughout the world.

ZOISITE, $Ca_2Al(Al_2O)(OH)(Si_2O_7)(SiO_4)$ (hydrous calcium aluminum silicate), is chemically identical to clinozoisite, but forms at higher temperatures and is different in structure. Also, it can take very little iron in aluminum positions without distortion. It is common in metamorphic rocks derived from basic igneous rocks and in metamorphosed sediments rich in calcium.

Zoisite crystallizes in orthorhombic system as prismatic, commonly columnar crystals and as masses. It is brittle, with perfect cleavage in one direction. Glassy in luster and generally transparent, it is gray, white, brown, green, or pink (thulite) and has no streak. It is not affected by acids. Hardness is 6-6.5, specific gravity 3.15-3.37. Very widespread in metamorphic rocks.

CLINOZOISITE, $Ca_2Al(Al_2O)(OH)(Si_2O_7)(SiO_4)$ (hydrous calcium aluminum silicate), accepts iron in aluminum positions and thus forms solid-solution series with epidote. Minerals with Al:Fe ratios of more than 9:1 are called clinozoisite; those with lower ratios, epidote. Clinozoisite is colorless, yellow, green, or pink. Other properties and occurrences are similar to epidotes.

PIEDMONTITE (or piemontite), $Ca_2(Mn,Fe,Al_3O(OH)(Si_2O_7)(SiO_4)$, has much Fe^{+3} and Mn^{+3} in Al^{+3} positions. Found only in Mn ores, as at Piedmonte, Italy, and in metamorphic rocks, as in France, Sweden, Japan. Also in quartz porphyries in Egypt and Adams Co., Pa. Deep red-brown or black. It is like epidote in crystal characteristics. Hardness is 6.5, specific gravity 3.40.

**EPIDOTE
CRYSTALS**
(Idealized)

Massive Epidote
Pershing, Colo.

Epidote Sprays
Baja California,
Mexico

**Iron-rich Epidote
with Feldspar**
North Carolina

Granular Epidote
Calumet, Colo.

Zoisite
Africa

Zoisite Crystals
Huntington, Mass.

**Iron-free Zoisite
with Quartz**
Timmins, Ontario

**Pink Zoisite
(Thulite)**
North Carolina

**Clinozoisite
Crystals**
Guerrero,
Mexico

Pink Clinozoisite
Sonora, Mexico

Piedmontite
Madera, Calif

161

GARNET GROUP is divided into two series of minerals. The pyralspite series is named after its three members, pyrope, almandite, and spessartite; the ugrandite series after uvarovite, grossularite, and andradite. An almost complete solid solution (p. 32) exists within each series, one member grading into another, but not between the two series. Garnets are among the commonest minerals. They crystallize in the cubic system, commonly forming dodecahedral and trapezohedral crystals in various combinations and less commonly rounded grains and granular masses. Garnets are brittle, but have no distinct cleavage. All are hard (6-7.5), have a glassy luster, are translucent to transparent, and have no streak. Their structure consists of independent silica tetrahedra, $(SiO_4)^{-4}$, held together by Ca^{+2}, Fe^{+2}, Mg^{+2}, or Mn^{+2} ions surrounded by eight O^{-2} ions and Al^{+3}, Fe^{+3}, or Cr^{+3} ions surrounded by six O^{-2} ions. They are found in a variety of metamorphic rocks and in granites, particularly pegmatites; hydrothermal pressure is important in their formation.

ALMANDITE, $Fe_3Al_2Si_3O_{12}$ (iron aluminum silicate), is the most abundant garnet. A deep red or brownish red in color, it is formed typically in schists in areas of regional metamorphism (p. 12). Large specimens are found at Gore Mt., N. Y. Almandite also occurs in a variety of igneous rocks in special cases and, because of its resistance to weathering, may be an important constituent of sedimentary rocks.

ANDRADITE, $Ca_3Fe_2Si_3O_{12}$ (calcium iron silicate), is very dark red, yellow, green, brown, or black. It commonly occurs in contact metamorphic rocks (p. 12) in impure limestones, notably in Sweden, W. Germany, U.S.S.R., and Franklin, N. J. It is found less commonly in calcium-rich rocks. Some varieties contain appreciable amounts of titanium.

GROSSULARITE, $Ca_3Al_2Si_3O_{12}$ (calcium aluminum silicate), is distinguished by its pale yellow or brown color, though it may also be pale green if it contains some chromium or rarely even reddish. It is found in metamorphosed impure limestones and limy shales, but only if the aluminum content is high and the iron content is low. It is more common in contact metamorphic deposits, notably in the Scottish Highlands.

PYROPE, $Mg_3Al_2Si_3O_{12}$ (magnesium aluminum silicate), is deep red, black, pink, or purple. Transparent varieties, notably from Czechoslovakia and North Carolina, are valuable as gems. Pyrope occurs with other ferromagnesian minerals in silicate-poor rocks such as the diamond pipes (kimberlites) of South Africa and in very high-grade metamorphic rocks. Rhodolite is a pink garnet similar to pyrope.

SPESSARTITE, $Mn_3Al_2Si_3O_{12}$ (manganese aluminum silicate), is dark red, violet, or brownish red. It is relatively rare, found most notably in Spessart district of Bavaria, West Germany; Italy; and Malagasy Republic. In these places it is associated with manganese ores of metamorphic origins. It may contain significant quantities of iron.

UVAROVITE, $Ca_3Cr_2Si_3O_{12}$ (calcium chromium silicate), is the rarest of all garnets. It is commonly bright emerald green and prized as a gem. The green color is characteristic of the Cr^{+3} ion. Uvarovite may contain some Al^{+3}. It is found associated with chromium ores in U.S.S.R., Quebec, and Spain, and with ferromagnesian rocks in Finland, Norway, and South Africa. It is metamorphic in origin.

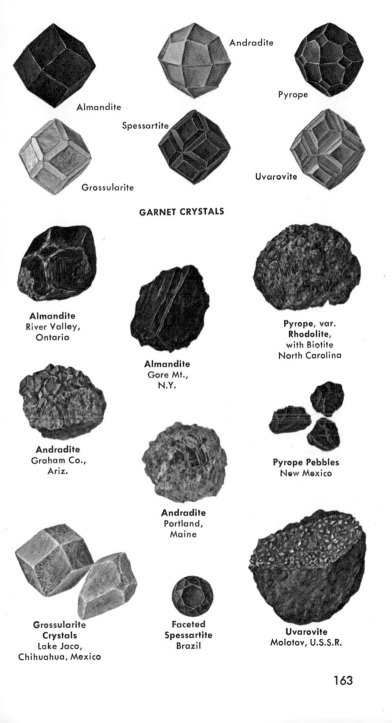

Almandite

Andradite

Pyrope

Spessartite

Grossularite

Uvarovite

GARNET CRYSTALS

Almandite
River Valley,
Ontario

Almandite
Gore Mt.,
N.Y.

Pyrope, var.
Rhodolite,
with Biotite
North Carolina

Andradite
Graham Co.,
Ariz.

Andradite
Portland,
Maine

Pyrope Pebbles
New Mexico

Grossularite
Crystals
Lake Jaco,
Chihuahua, Mexico

Faceted
Spessartite
Brazil

Uvarovite
Molotov, U.S.S.R.

OLIVINE GROUP includes minerals whose structure consists of silica tetrahedra, $(SiO_4)^{-4}$, held together by magnesium, Mg^{+2}, or certain other ions in octahedral positions (p. 156). Only half of the octahedral positions are filled by these ions, however, and only an eighth of the tetrahedral positions are filled by silicon ions, Si^{+4}. Olivine-group minerals crystallize in the orthorhombic system as granular masses and scattered grains, rarely as crystals. They are brittle, fracturing irregularly; are translucent to transparent, with a glassy luster; and have no streak. They are soluble in hydrochloric acid. Olivine is the name given to any of the minerals that form a complete solid solution series (p. 32) ranging from forsterite at one end to fayalite at the other. The minerals in-between contain varying proportions of iron, Fe, and magnesium, Mg, and have the general formula $(Fe,Mg)_2 SiO_4$. Specimens of either pure end member are rare, but the olivines in-between are common. Magnesium-rich olivines are commonest because magnesium is more abundant than iron. The series is one of the major rock-forming groups and may hold the key to an understanding of the earth's interior and of mountain-building processes. Olivine-rich rocks, called dunites and peridotites, are common in zones of high shearing stress on the flanks of folds in strongly folded mountain belts. The diamond pipes of South Africa are peridotites, with olivine strongly altered to serpentine in the older volcanic rocks.

FORSTERITE, Mg_2SiO_4 (magnesium silicate), is pale green to dark green, readily weathering to reddish brown. Hardness is 6.5-7, specific gravity 3.27. It is found throughout the world as scattered grains in basic (silica-poor) rocks, as massive intrusions in mountain cores, and as major constituents of magnesium-rich metamorphic rocks. Peridot, the gem variety, is light green and transparent.

FAYALITE, Fe_2SiO_4 (iron silicate), is generally dark green to black, weathering rapidly to rust as the iron oxidizes. Hardness is 6.5, specific gravity 4.1. It is most common in metamorphosed iron-rich sediments, as in Harz Mts., West Germany. It may occur in deep-seated intrusive rocks, as in Greenland, and in smaller amounts in volcanic rocks. Also found in Sicily, Azores, France, Sweden, Ireland, Massachusetts, Colorado, California, and Wyoming. If it contains much manganese it is called manganfayalite.

TEPHROITE, Mn_2SiO_4 (manganese silicate) is a rare mineral that occurs in small red or gray crystals. Hardness is 6, specific gravity 4.1. It is found in manganese-rich rocks at Franklin, N. J., in Vermland, Sweden, and in the French Pyrenees. A zinc-containing variety is called roepperite. Specimens containing enough iron so that the composition approximates $FeMnSiO_4$ are called knebelite. May contain appreciable magnesium.

MONTICELLITE, $CaMgSiO_4$ (calcium magnesium silicate), occurs as small prismatic crystals, rounded grains, or masses that are colorless or gray. The Ca:Mg ratio is very near 1:1 in all cases. It is softer than other minerals of the group (hardness: 5), and its specific gravity is 3.2. It forms normally in metamorphic rocks derived from siliceous dolomites, SiO_2 and $CaMg(CO_3)_2$, or under unusual conditions in calcium-rich igneous rocks. It is found in Italy (at Mt. Vesuvius), Australia, California, and Arkansas.

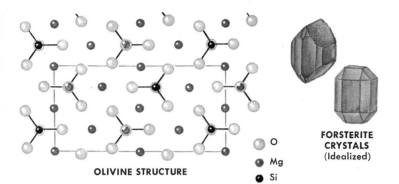

O

Mg

Si

FORSTERITE CRYSTALS
(Idealized)

OLIVINE STRUCTURE

Massive Forsterite
Jackson Co., N.C.

Forsterite
Crestmore, Calif.

Peridot, Gem Quality Olivine
Arizona

**Tephroite (Tan)
with Franklinite**
Franklin Furnace, N.J.

Fayalite with Cristobalite
Inyokern, Calif.

Massive Monticellite
Crestmore, Calif.

Monticellite with Blue Calcite
Crestmore, Calif.

SUBSATURATE GROUP includes independent tetrahedral silicates that have in their structure oxygen ions in addition to those forming the tetrahedra. The formulae, then, are written in such a way as to distinguish between the two kinds of oxygen. Technically any silicate containing the hydroxyl ion, $(OH)^{-1}$, is a subsaturate, but these are normally not so classified.

KYANITE, Al_2OSiO_4 (aluminum silicate), is found exclusively in regional metamorphic rocks rich in aluminum oxide, where it occurs with micas (in mica schists), staurolite, garnet, or corundum. It is very common in the Piedmont regions of North Carolina, in the Alps, and in the Urals. It generally forms at temperatures higher than staurolite but lower than sillimanite. Kyanite crystallizes in triclinic system as long, bladed crystals or as masses, usually distinctively blue in color. It may also be white, gray, green, brown, or black. It is brittle, with cleavage in two directions not at right angles. Glassy in luster, it is translucent to transparent and has no streak. Hardness is 5-7, sometimes varying from crystal face to face; specific gravity 3.53-3.65. Because of its high melting point, it is used as a raw material for refractories.

STAUROLITE, $(Fe,Mg)_2(Al,Fe)_9O_6(SiO_4)_4(O,OH)_2$ (iron-magnesium aluminum silicate), is commonly found with garnet, mica, and quartz in areas of regional metamorphism. It is abundant in mica schists along the Appalachian chain from New England to Georgia and in many other geologically similar areas. Composition varies, with wide differences in Fe:Mg and Al:Fe ratios in different specimens; much Mn is found in staurolites in Sweden. Staurolite crystallizes in orthorhombic system as prismatic and flattened crystals, cruciform twins being typical. It is brittle, with cleavage in one direction. Generally opaque because of impurities, it is brown, black, or yellow, with resinous or glassy luster, and may have pale gray streak. Hardness is 7-7.5, specific gravity 3.74-3.83.

SILLIMANITE, Al_2OSiO_4 (aluminum silicate), has the same composition as kyanite, but a different structure and characteristics. It occurs in silicate deficient metamorphic rocks as needlelike crystals in orthorhombic system and as fibrous groups. It is a glassy brown, white, or green. Cleavage occurs in only one direction; hardness 6-7, specific gravity 3.23-3.27.

ANDALUSITE, Al_2OSiO_4 (aluminum silicate), is chemically identical to kyanite and sillimanite, but probably forms at lower temperature. It is found in metamorphic regions, partly altered to kyanite. It crystallizes in orthorhombic system as coarse prisms and masses—glassy white, red, gray, brown, or green. Cleavage is in two nearly perpendicular directions. Hardness is 7.5, specific gravity 3.13-3.16.

TOPAZ, $Al_2(SiO_4)(OH,F)_2$ (aluminum fluosilicate), is not related to the other minerals here, but does have a similar structure. It occurs in granites, particularly pegmatites, commonly associated with beryl, tourmaline, fluorite, and other pegmatite minerals and with tin and tungsten ores. Transparent topaz is often used as a gemstone. It is very durable and has a high index of refraction. Topaz crystallizes in orthorhombic system as columnar prisms, commonly with striated faces, and as granular masses. It is brittle, with perfect cleavage in one direction. It is transparent, glassy in luster, and colorless, pale streak). Hardness is 8, specific gravity 3.49-3.57. Occurs in U.S.S.R., East Germany, Nigeria, Japan, Maine, Connecticut, New Hampshire, Texas, Virginia, Utah, California, etc.

Massive Kyanite
Lincolnton, N.C.

Photomicrograph of Sillimanite in Calcite,
Inwood Marble
Pound Ridge, N.Y.

North
Carolina

Minas Gerais,
Brazil

Bladed Kyanite Crystals

Raw Stones

Faceted

Andalusite
Brazil

Sillimanite
Hill City, S. Dak.

Massive Andalusite
with Pyrophyllite
Hemet, Calif.

Andalusite,
var. Chiastolite,
with Mica Coating
Madera Co., Calif.

Topaz

Topaz
Pitkin, Colo.

Citrine
Topaz
Brazil

Topaz
Gunnison,
Colo.

Staurolite Crystals
Fannin Co., Ga.

Topaz
Villa Rica,
Brazil

Topaz
San Luis Potosi,
Mexico

Topaz
Mexico

167

DOUBLE TETRAHEDRAL SILICATES

These consist of double tetrahedral units, $Si_2O_7)^{-6}$, each made up of two tetrahedra, sharing one oxygen ion, which is saturated. The remaining six oxygen ions, which carry one negative charge each, must be balanced by the positively charged ions that hold the crystal together. Because the $(Si_2O_7)^{-6}$ unit has an unusual shape, the structures are complex. Double tetrahedral silicates are relatively uncommon, being represented in this book by the melilite group and by the epidote group (p. 162), which has both independent and double tetrahedral units.

MELILITE GROUP includes a complete solid solution series of minerals (p. 32) between gehlenite and akermanite, intermediate minerals being called melilites. In the melilite structure, as illustrated by akermanite, double tetrahedra are held together by Mg^{+2} ions arranged tetrahedrally in sheets held together by bonds between Ca^{+2} and O^{-2} ions. An Al^{+3} ion can replace an Mg^{+2} ion if at the same time an Al^{+3} replaces an Si^{+4}. The end result of such substitutions is gehlenite. Melilites are characteristic of metamorphosed impure limestones formed at high temperatures, particularly in contact zones (p. 12). Sodium and iron are common constituents.

GEHLENITE, $Ca_2Al(AlSiO_7)$ (calcium aluminum silicate) forms in rocks with a high aluminum and low magnesium content. It is commonly found in limestones invaded by silica-deficient (basic) igneous material and in basaltic lavas.

 Gehlenite crystallizes in tetragonal system as square prisms and plates and as masses. It is white, yellow, green, red, or brown, and may give a pale streak. Opaque to transparent, it has a glassy or resinous luster and is soluble in hydrochloric acid. Cleavage is distinct in one direction. Hardness is 5, specific gravity 3.04. Occurs in Italy, Mexico, Rumania, South Africa, Quebec, U.S.S.R., Colorado, etc.

AKERMANITE, $Ca_2Mg(Si_2O_7)$ (calcium magnesium silicate), is relatively uncommon in the pure state, most melilites containing significant quantities of Al. It is often found in industrial slags, in which it normally contains abundant iron. At low temperatures it breaks down and forms monticellite and wollastonite. This is probably the result of structural instability of the Mg^{+2} ion in tetrahedral coordinations. Properties are similar to those of gehlenite.

HARDYSTONITE, $Ca_2Zn(Si_2O_7)$ (calcium zinc silicate), is the zinc analog of akermanite, with Zn^{+2} occupying Mg^{+2} positions. An appreciable quantity of manganese, Mn^{+2}, is also present, probably in Ca^{+2} positions. Hardystonite has been found only in the zinc mines of Franklin, N. J. Properties are very similar to those of other melilites except that hardystonite occurs in white granular masses and has a hardness of 3-4 and a specific gravity of 3.4.

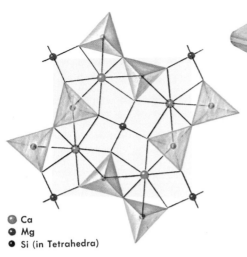

- Ca
- Mg
- Si (in Tetrahedra)

MELILITE STRUCTURE

MELILITE CRYSTALS
(Idealized)

Melilite,
var. Somervillite
Mt. Vesuvius, Italy

Melilite, var. Fuggerite
Austria

Melilite, var. Humboldtilite
Mt. Vesuvius, Italy

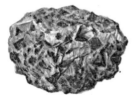

Gehlenite
Austria

Gehlenite (Massive)
Luna Co., N. Mex.

Akermanite
Trentino, Italy

Hardystonite
Franklin, N.J.

RING SILICATES

Ring silicates are structurally complex and interesting minerals. They are relatively uncommon because of the chemical conditions necessary for their growth; chain silicates, chemically similar to ring silicates, are more readily formed because they consist of more abundant elements, and are chemically more stable. A silicate ring is formed when each silica tetrahedron shares two oxygen ions with adjacent tetrahedra. The ring thus consists of a number of $(SiO_3)^{-2}$ units. Rings are held together by positive ions coordinated with the oxygen ions of the rings. Because the bonds between these positive ions and the oxygen ions are weak and distributed rather regularly in all directions, ring silicates do not exhibit well-developed cleavage. The geometry of the ring, however, is commonly exhibited by the external form of the crystal.

The most common ring structures have six tetrahedra per ring. These exhibit hexagonal or trigonal symmetry, depending on the shape of the ring. Tourmaline, beryl, cordierite, and the less common osumilite are of this type. Benitoite has a three-membered ring, axinite four-membered. Only the first three are described.

TOURMALINE, $Na(Mg,Fe,Mn,Li,Al)_3Al_6(Si_6O_{18})(BO_3)_3 \cdot (OH,F)_4$ (sodium aluminum borosilicate) is commonly used as a gemstone. It consists of six-membered silica rings and three-membered borate rings held together by sodium, aluminum, and other positive ions. The three main varieties are dravite, a brown magnesium-rich tourmaline; schorl, a very black and opaque iron- and manganese-rich tourmaline; and elbaite, a blue, green, yellow, red, white, or colorless lithium-rich tourmaline. Some trigonal prisms are prized as gemstones for the color-banding caused by changes in composition during crystal growth.

Tourmalines are common as late-stage crystallization products of granites, particularly pegmatites, that have been enriched by boron-bearing solutions. They also form in some metamorphic rocks and appear as scattered, rounded grains in sediments. They commonly occur as radiating clusters and as isolated crystals with quartz, muscovite, and feldspars. Crystals are typically trigonal prisms with round, striated faces.

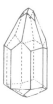

Tourmaline crystallizes in hexagonal system as prisms, radiating aggregates, and masses; twinning is rare. Opaque to transparent, it has glassy luster and no streak; it is insoluble in acids and is pyro-electric (becomes electrically charged when heated or cooled). Hardness is 7, specific gravity 3.03-3.25. Exceptional crystals come from U.S.S.R., Island of Elba, Greenland, West Germany, Czechoslovakia, Switzerland, England, Ceylon, Brazil, New England (particularly West Paris, Me.), and California (San Diego area).

Packing Model

IDEAL 6-MEMBERED RING $(Si_6O_{18})^{-12}$

Tetrahedral Model

Packing Model

Tetrahedral Model

IDEAL 3-MEMBERED RING $(Si_3O_9)^{-6}$

Stubby Prism

Flat Prism

Slender Prism

Slender Prism

3-fold Symmetry

TOURMALINE CRYSTALS

Black Tourmaline (Schorl) Pierrepont, N.Y.

Violet Tourmaline Pala, Calif.

Tourmaline in Feldspar North Carolina

Rubellite Tourmaline in Lepidolite Mica Pala, Calif.

Green Tourmaline Brazil

Tourmaline in Quartz Brazil

171

BERYL, $Be_3Al_2Si_6O_{18}$ (beryllium aluminum silicate), has a structure consisting of six-membered rings arranged in sheets. Beryllium and aluminum ions between sheets hold together rings within sheets and rings in adjacent sheets. Rings in different sheets are aligned to form channels. In these channels sodium, lithium, and cesium are commonly found.

Beryl is found in granite, particularly in pegmatites, associated with quartz, feldspar, muscovite, topaz, tourmaline, lepidolite, spodumene, and other pegmatite minerals. Well-formed crystals of huge size (up to 200 tons) are common. It also occurs in silica-poor schists and marbles. Clear varieties of beryl, such as emerald (green), aquamarine (blue-green), and heliodor (yellow), are highly prized as gemstones. In recent years beryl also has become the major source of beryllium metal and beryllium oxide, both important industrial materials. Beryl is recognized by its typical crystal form (six-sided prism), its color (usually pale green or emerald green, sometimes blue-green, yellow, pale red, white, or colorless), and its hardness (7.5-8).

Beryl crystallizes in hexagonal system as long prisms (commonly striated, seldom twinned), granular masses, and coarse columns. It is translucent to transparent, and glassy in luster; has no cleavage and no streak. It may contain alkalies and water, and some magnesium, calcium, chromium (in emeralds), and other elements. Widely distributed in small amounts, it occurs in New England; Smoky Mts., N.C.; Colorado; San Diego Co., Calif.; U.S.S.R.; Austria; Malagasy Rep; Columbia; etc.

CORDIERITE, $Mg_2Al_4Si_5O_{18}$ (magnesium aluminum silicate), has essentially the same structure as beryl, but its different composition distorts the rings into a normally orthorhombic structure (a high-temperature hexagonal type is also known). Cordierite is commonest in metamorphic rocks, forming at high temperatures from magnesium- and aluminum-bearing sediments, and less common in volcanic rocks. It closely resembles quartz, but is commonly altered by weathering, especially along fractures, to grayish-green mica minerals. Because cordierite has a low thermal expansion coefficient, it does not fracture easily during rapid heating or cooling. Refractories, have thus been made by reacting talc and kaolinite to form cordierite.

Cordierite crystallizes in orthorhombic system usually as grainy masses, sometimes as short single or twinned prisms. It is gray, blue, or smoky, with no streak; translucent to transparent; glassy to greasy in luster; and brittle, with fair cleavage in one direction. It forms a complete solid solution with iron cordierite ($Fe_2Al_4Si_5O_{18}$), contains some calcium and water, and is partially decomposed by acids. Hardness is 7-7.5, specific gravity 2.53-2.78. Occurs chiefly in Finland, Norway, Greenland, W. Germany, Malagasy Rep., Connecticut, and Colorado.

BERYL STRUCTURE

Al

Be

SiO₄ } Tetrahedra

BERYL CRYSTALS
(Idealized)

Beryl Crystal
Grafton, N.H.

Beryl Crystals
Brazil

Beryl Crystal
Afghanistan

Beryl Crystal
Deshong's Quarry,
Delaware Co., Pa.

Beryl (Heliodor)
Topsham, Maine

Aquamarine
Brazil

Cordierite
Norwich, Conn.

Cordierite
Malagasy Republic

**CORDIERITE
CRYSTAL**
(Idealized)

SINGLE-CHAIN SILICATES

The single-chain silicates are a major group of rock-forming minerals and are therefore of great interest to mineralogists and petrologists. They consist of single chains of silica tetrahedra. Each tetrahedron in a chain shares two of its oxygen ions, O^{-2}, with adjacent tetrahedra in the chain. The shared oxygen ions, then, link the tetrahedra together. This sharing reduces the number of oxygen ions per unit by one. The result is a series of $(SiO_3)^{-2}$ units. The two negative charges per unit are carried on the two unshared oxygen ions. These charges must be balanced in the crystal by positive ions—most commonly magnesium, Mg^{+2}; iron, Fe^{+2}; and calcium, Ca^{+2}—located between chains. Because the bonds between chains are weaker than the silicon-oxygen-silicon bonds within chains, cleavage occurs in directions corresponding to the long axes of the chains.

The chains may be packed together in many different ways, depending on the various sizes and charges of the ions between chains. As a result, single-chain silicates are more abundant than ring silicates and have wide variations in composition. Though the single chain is commonly pictured as straight, the chains in real crystal structures are commonly twisted or helical to accommodate ions of different sizes between them.

PYROXENE GROUP includes all single-chain silicates, and the latter are generally referred to as pyroxenes. Because Mg^{+2}, Fe^{+2}, and Ca^{+2} are the commonest metallic ions of suitable size to occupy positions between single chains of silica tetrahedra, pyroxenes containing combinations of these elements are most important as rock-forming minerals. The pyroxenes shown on the triaxial diagram on the opposite page are the commonest types. Wollastonite contains little Mg^{+2} or Fe^{+2} and is generally not associated with the other pyroxenes. Diopside and hedenbergite form a complete solid-solution series (p. 32) with diopside as the Mg-rich end member and hedenbergite as the Fe-rich end member. The intermediate member, augite, contains all three metals—Mg, Fe, and Ca. Enstatite and ferrosilite are the end members of another complete solid-solution series. All can contain a little Ca, and the intermediate member, hypersthene, contains both Mg and Fe and some Ca. Ferrosilite is very rare.

In addition to these rock-forming minerals, aegirite, $NaFeSi_2O_6$; jadeite, $NaAlSi_2O_6$; and spodumene, $LiAlSi_2O_6$, are locally abundant and of great geological significance. Rhodonite, $MnSiO_3$, and johannsenite, $Ca(Mn,Fe)Si_2O_6$, are interesting pyroxenes found in Mn-bearing deposits. Many other names have been applied to compositional variations of the common pyroxenes.

Pyroxenes of many compositions form in industrial processes, notably in slags in steel-making. A mixture of enstatite ($MgSiO_3$) and forsterite (Mg_2SiO_4) forms when serpentine—a common component of furnace linings—is heated.

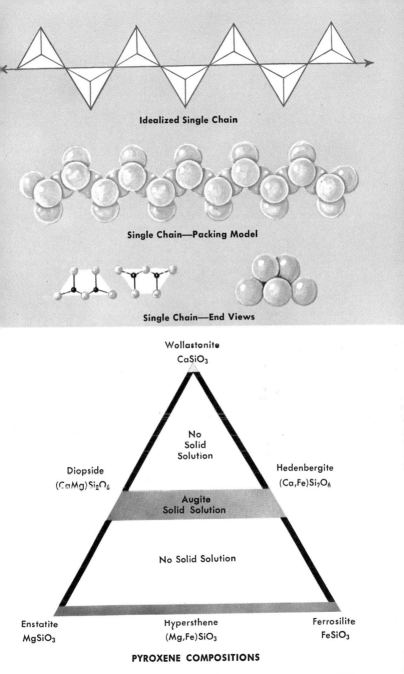

Idealized Single Chain

Single Chain—Packing Model

Single Chain—End Views

Wollastonite
$CaSiO_3$

No
Solid
Solution

Diopside
$(CaMg)Si_2O_6$

Hedenbergite
$(Ca,Fe)Si_2O_6$

Augite
Solid Solution

No Solid Solution

Enstatite
$MgSiO_3$

Hypersthene
$(Mg,Fe)SiO_3$

Ferrosilite
$FeSiO_3$

PYROXENE COMPOSITIONS

175

DIOPSIDE, $CaMgSi_2O_6$ (calcium magnesium silicate), is commonly formed by metamorphism of siliceous dolomite, $CaMg(CO_3)_2$, at relatively high temperatures. The pure compound is white, but small amounts of iron are usually present, producing a greenish or yellowish color. Diopside is not common as an igneous mineral (augite is the common igneous pyroxene). Diopside commonly occurs as distinctive, small, rounded crystals in coarse marbles, where it is associated with phlogopite, serpentine, and other magnesium silicates. Other names applied to varieties include alalite, malacolite, violan, canaanite, and lavrovite.

Diopside crystallizes in monoclinic system in form of granular, prismatic crystals, scattered grains, masses, and columns. Twin and multiple crystals are common. It is white, yellowish, green, or blue; glassy; translucent to transparent; brittle. Has no streak. It cleaves in two directions at 87° (angle distinguishes it from amphiboles). It is insoluble in acids; weathers easily. Very common in areas of contact metamorphism in Appalachians, Rockies, Alps, Urals, Sweden, Canada, etc.

ENSTATITE, $MgSiO_3$ (magnesium silicate), is the simplest pyroxene in composition. It occurs only in magnesium-rich rocks that have little calcium or iron. These include silica-poor, deep-seated igneous rocks, lavas, and metamorphic rocks of igneous origin. (Igneous rocks of intermediate silica content contain amphiboles or biotite rather than pyroxenes.) Enstatite is difficult to interpret geologically because the silica chains of its structure are linked together in three basically different ways, depending on conditions of formation, and because calcium and iron in its structure cause variations in its properties.

Enstatite crystallizes in orthorhombic or monoclinic system as interlocking grains or prismatic crystals. Repeated parallel twinning is common. Mineral is colorless, gray, yellow, green, or brown; dull, glassy, pearly, or bronze-like in luster; translucent; brittle; and insoluble in acids. It cleaves in 2 directions at 88° angle, fractures unevenly. May be difficult to tell from augite. Hardness is 5-6, specific gravity 3.2 (higher as iron content increases). Enstatite is widespread in igneous rocks and common in meteorites.

HEDENBERGITE, $CaFeSi_2O_6$ (calcium iron silicate), is uncommon because most rocks with high iron content also have high magnesium content. It is found in some gabbros and other basic igneous rocks, with lead and zinc ores in contact metamorphosed limestones, and as a metamorphic product in iron-rich sedimentary rocks. It crystallizes in monoclinic system as prismatic crystals, granular masses, and lamellar aggregates. Twinning is common. Color is brownish green, dark green, or black. Hedenbergite cleaves in 2 directions at 87° angle. It commonly contains Mg and other elements, and is insoluble in acids. It is common in Sweden, Norway, Siberia, New York, and other scattered localities.

PYROXENE CLEAVAGE
(Idealized)

DIOPSIDE CRYSTALS
(Idealized)

Crystal Group

Crystalline

Diopside, Gasconade, Mo.

Diopside with Biotite Quebec

Enstatite Delaware Co., Pa.

Enstatite (ult. to Steatite) Brevik, Norway

Enstatite Tilly Foster Mine, Brewster, N.Y.

Enstatite, var. Bronzite Lancaster Co., Pa.

Hedenbergite Silverstar, Mont.

AUGITE, $(Ca,Mg,Fe,Ti,Al)_2(Si,Al)_2O_6$ (calcium-magnesium-iron-titanium-aluminum silicate), is the commonest igneous pyroxene; its complex chemistry reflects the complexity of igneous melts. It is formed at high temperatures early in the crystallization process. Augite is common in silica-poor (basic) igneous rocks, notably gabbros and basaltic lavas. It occurs as scattered crystals, evenly distributed throughout the rock; some basic intrusive rocks are nearly all augite. It is uncommon in metamorphic rocks of sedimentary origin. Augite is difficult to distinguish from other dark pyroxenes because all have similar structures. A chemical analysis is needed to characterize most pyroxenes.

Augite crystallizes in monoclinic system as prismatic crystals (often twinned) and granular masses. It is a glassy brown, green, or black; translucent to opaque; brittle; and insoluble in acids. Cleavage is in 2 directions at 87°, fracture uneven. Hardness is 5.5-6, specific gravity 3.23-3.52. Augite occurs in lavas of Vesuvius, Stromboli, Hawaii, and Czechoslovakia; at Ducktown, Tenn., and Franklin, N. J.; and in New York, Connecticut, Massachusetts, Utah, Canada, and elsewhere.

HYPERSTHENE, $(Mg,Fe)SiO_3$ (magnesium-iron silicate), occurs in dark granular igneous rocks and in some meteorites. An intermediate member of the enstatite-ferrosilite series, its Mg:Fe ratio varies widely. It crystallizes in orthorhombic system as prisms or tabular masses. It is dark brownish green or black, with pearly luster; brittle, and easily weathered. Cleavage is more prominent in one direction than the other, fracture uneven. Hardness is 5-6, specific gravity 3.40-3.50.

AEGERITE, $NaFeSi_2O_6$ (sodium iron silicate), is common in syenites. Composition varies, with calcium-magnesium varieties called aegerite-augite. Some other varieties are called acmite. A rare mineral, it occurs most notably in Magnet Cove, Ark.; Bear Paw Mts., Mont.; Norway, and Greenland. It crystallizes in monoclinic system as slender prismatic crystals and fibrous masses. It is a glassy, translucent brown, or green. Hardness is 6-6.5, specific gravity 3.40-3.55.

JOHANNSENITE, $Ca(Mn,Fe)Si_2O_6$ (calcium manganese-iron silicate), is relatively rare. It occurs in limestones that were affected by Mn-rich solutions during metamorphism, and in veins. Mn and Fe content varies. Mineral crystallizes in monoclinic system as fibrous masses or prismatic crystal aggregates. It is brown, grayish, or green, commonly with black surface stains. Cleavage is in 2 directions at 87°. Hardness is 6, specific gravity 3.44-3.55. Mineral is common in manganese deposits in Italy, Mexico, and Australia.

RHODONITE, $MnSiO_3$ (manganese silicate), has a distinctive pink or red color (similar to that of rhodochrosite, $MnCO_3$) because of its high manganese content. It always contains appreciable calcium. It crystallizes in the triclinic system as tabular crystals, masses, and scattered grains. Brown or black oxidation products are common on its surface. Cleavage is in 2 directions at 92.5°. Hardness is 5.5-6.5, specific gravity 3.57-3.76. Rhodonite is found in manganese deposits, as at Franklin, N. J., and in metamorphic rocks.

178

AUGITE CRYSTALS
(Idealized)

Augite Crystals
Tanzania

Pyroxene
Sonoma Co.,
Calif.

**Pyroxene with
Wollastonite**
Glacial Erratic,
Ohio

Augite Crystal
Otter Lake,
Quebec

Hypersthene
Lake St. John,
Quebec

Aegerite
Narsarssuak,
Greenland

**Johannsenite
with Rhodonite**
California

**Rhodonite
with Quartz**
Butte, Ont.

**Rhodonite
with Quartz**
Australia

Rhodonite, var. Fowlerite
Buckwheat Mine,
Franklin, N.J.

JADEITE, NaAl(SiO₃)₂ (sodium aluminum silicate), is one of two minerals (nephrite is the other) that when cut become the ornamental stone and precious gem, jade. It occurs typically as a dense, massive rock suitable for carving large art objects. Jadeite is noted for its toughness and density, a consequence of its formation at high pressure. Found in metamorphic environments with other high-pressure pyroxenes and with albite, it is intermediate in composition between albite, $NaAlSi_3O_8$, and silica, SiO_2.

Jadeite crystallizes in monoclinic system as granular or foliated masses. It is colorless, shades of green, or black; color is due to iron content. Luster is somewhat glassy. Cleavage is prismatic, fracture splintery. Hardness is 6.5-7, specific gravity 3.3-3.5. Commonly found among artifacts of early man. Occurs chiefly in metamorphic areas of China, Burma, Sulawesi, Guatemala, Mexico, Japan, and California.

WOLLASTONITE, CaSiO₃ (calcium silicate), forms exclusively at high temperatures by the reaction of $CaCO_3$ and SiO_2 in contact or regional metamorphic areas. If much iron or magnesium is present, however, the commoner pyroxenes, augite, diopside, and hedenbergite, will form; thus the formation of wollastonite is generally restricted to the reaction of large bodies of $CaCO_3$ (limestone or marble) with quartz or silica-rich solutions.

Wollastonite crystallizes in monoclinic system as tabular or short prismatic crystals or as granular masses. It is white to gray, yellow, red, or brown; glassy or pearly in luster; translucent; and brittle. Cleavage is platy or fibrous, fracture uneven. Hardness is 4-4.5, specific gravity 2.8-2.9. Found in New York, Michigan, California, Quebec (Grenville), Rumania, Mexico, and Italy (Vesuvius).

SPODUMENE, LiAl(SiO₃)₂ (lithium aluminum silicate), is the chief source of lithium. It crystallizes exclusively in Li-rich pegmatites, occurring with quartz, microcline, albite, beryl, and the lithium minerals lepidolite and rubellite tourmaline. It is typically dull gray or pearly white, but emerald green variety, hiddenite and lilac variety, kunzite, are highly prized as gemstones. It is brittle and transparent to translucent; occurs as prismatic crystals (monoclinic system) or as masses. Some crystals weigh many tons. Occurs in Appalachian Mts. from Maine to North Carolina; also in South Dakota, California, Sweden, Iceland, Brazil, and Malagasy Rep.

CHRYSOCOLLA, CuSiO₃·nH₂O (hydrous copper silicate), is included here because of its reported composition. Structure is not known, but is probably amorphous. Chrysolla is an alteration product of copper minerals. It is commonly mixed with hydrated copper silicates, bisbeeite and shattuckite, in the upper levels of vein deposits, as in Cornwall, England; Katanga Prov., Congo; Chile; and Arizona. It is opal-like, with banded and botryoidal forms common, but is too soft (2-2.24) to be useful as a gemstone. It occurs as opaque-to-translucent crusts or seam fillings in shades of blue and green, black, or brown, with inclusions. Streak is white.

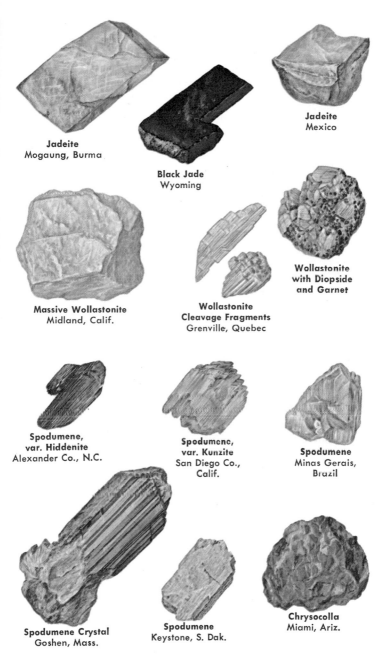

Jadeite
Mogaung, Burma

Black Jade
Wyoming

Jadeite
Mexico

Massive Wollastonite
Midland, Calif.

**Wollastonite
Cleavage Fragments**
Grenville, Quebec

**Wollastonite
with Diopside
and Garnet**

**Spodumene,
var. Hiddenite**
Alexander Co., N.C.

**Spodumene,
var. Kunzite**
San Diego Co.,
Calif.

Spodumene
Minas Gerais,
Brazil

Spodumene Crystal
Goshen, Mass.

Spodumene
Keystone, S. Dak.

Chrysocolla
Miami, Ariz.

DOUBLE-CHAIN SILICATES

The basic structural unit of double-chain silicates is an infinite chain (p. 159) consisting of two single chains that share oxygen ions. The residual charge thus is on the apices of the tetrahedra and on the unshared edges. Each of the hexagonal holes formed by the linkage of the single chains into a double chain must be occupied by a hydroxyl ion, $(OH)^{-1}$, or the unit is unstable. The resulting unit, $[(Si_4O_{11})OH]^{-7}$, is neutralized by positive ions located between chains. Because many different ions may be present and many ionic substitutions are possible, the chemical variability of these silicates is remarkable. The chemical relationships of common amphiboles are listed on the next page.

AMPHIBOLE GROUP includes all double-chain silicates. Because amphiboles contain water in their structure in the form of OH ions, they are formed in igneous rocks at lower temperatures than olivines or pyroxenes; at higher temperatures they lose water and break down to pyroxenes and olivines. Amphiboles are also formed in metamorphic rocks. Ions in various combinations may be found in structural positions between chains; aluminum ions, Al^{+3}, may substitute for silica, Si^{+4}, in tetrahedral positions, and fluorine, F^{-1}, may substitute for $(OH)^{-1}$. The resulting crystals may be monoclinic or orthorhombic, and solid solution (p. 32) is the rule. Of the more than 30 amphiboles, only a few common ones are described.

TREMOLITE, $Ca_2Mg_5(Si_8O_{22})(OH,F)_2$ (calcium magnesium hydrous silicate), is the magnesium-rich end member of a solid-solution series; the other end member, actinolite, is rich in iron. Most amphiboles in the series contain both magnesium and iron and are called tremolite-actinolite. They are metamorphic in origin, resulting from alteration of siliceous dolomites; thus they are common in marbles.

Tremolite crystallizes in monoclinic system as bladed crystals, fibrous or granular masses; a compact variety is called nephrite. It is colorless, or gray or brown to black (becoming greener as iron content increases). It is brittle, glassy to dull in luster, transparent to opaque. Typical cleavage is about 56°, fracture uneven. Hardness is 5-6, specific gravity 3-3.44. It is found in many metamorphic areas, such as New England, Japan, Great Britain, and New Zealand.

ACTINOLITE, $CaFe_5(Si_8O_{22})(OH,F)_2$ (calcium iron hydrous silicate), is dark green to black, but is very similar to tremolite in other respects. Nephrite jade is most commonly actinolite occurring as massive, tough, granular aggregates with other metamorphic minerals. Many varieties of asbestos, such as mountain leather, byssolite, and mountain wood, are tremolite-actinolite minerals with fibrous habit.

COMPOSITIONS OF THE AMPHIBOLES are both complex and variable because of extensive solid solution. A general formula, $A_{7-8}(Si, Al)_8O_{22}(OH, F, O)_2$, is applicable. The more common amphiboles, with the A elements are listed here.

ORTHORHOMBIC AMPHIBOLES
Anthophyllite   A = Mg, Fe
Gedrite-Ferrogedrite   A = Mg, Fe, Al
Holmquistite   A = Li, Mg, Fe, Al

MONOCLINIC AMPHIBOLES
Tremolite-Actinolite   A = Ca, Mg, Fe
*Hornblende   A = Ca, Na, K, Mg, Fe, Al
Kaersutite   A = Ca, Na, K, Mg, Fe, Ti
Barkevikite   A = Ca, Na, K, Mg, Fe, Mn
Glaucophane   A = Na, Mg, Al
Riebeckite   A = Na, Fe
Magnesioriebeckite   A = Na, Fe, Mg
Katophorite   A = Na, Ca, Fe, Al
Magnesiokatophorite   A = Na, Ca, Mg, Fe, Al
Eckermannite   A = Na, Ca, Mg, Fe, Al, Li
Arfvedsonite   A = Na, Ca, Fe, Mg, Al

* Other compositional varieties of hornblende are edenite, ferroedenite, tschermakite, ferrotschermakite, and ferrohastingsite

Fibrous

Granular (enlarged)

Tremolite, Balmat, N.Y.

Brown Tremolite
St. Lawrence Co., N.Y.

Massive Actinolite
Cardwell, Mont.

Bladed Actinolite
Chester, Vt.

Bladed Actinolite
Delaware Co., Pa.

HORNBLENDE, $(Ca,Na,K)_{2-3}(Mg,Fe,Al)_5(Si,Al)_8O_{22}(OH,F)_2$ (hydrous calcium-sodium-potassium magnesium-iron-aluminum silicate), is a perfect example of the chemical variability generally found in silicates. Small amounts of ions other than those in the formula may also be present. The composition of hornblendes varies widely, depending on chemicals available at formation and on the temperature of formation. Hornblende is associated with the other major rock-forming minerals in igneous and metamorphic rocks in all mountainous areas of the world and in basaltic lavas wherever they are found. Some large rock bodies consist of nearly all hornblende.

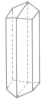

Hornblende crystallizes in monoclinic system as fibrous, stubby prismatic crystals. It is glassy green-black, generally opaque. Cleavage is in 2 directions at 56°, fracture uneven. Hardness is 5-6, specific gravity 3-3.5. May contain titanium, manganese, other elements.

ANTHOPHYLLITE, $(Mg,Fe)_7(Si_8O_{22})(OH,F)_2$ (hydrous magnesium-iron silicate), forms a solid-solution series with gedrite. It occurs in metamorphic environments where limestones are rare, and is commonly associated with cordierite (p. 174). It crystallizes in orthorhombic system as prismatic crystals (rare), asbestos fibers, and lamellar masses. It is white, gray, green, or brown, depending on iron content, and translucent to transparent. Cleavage is in 2 directions at 54½°, hardness 5.5-6, specific gravity 2.85-3.57. Notable deposits in North Carolina, Pennsylvania, Greenland, Norway, Austria, and Czechoslovakia.

GEDRITE, $(Mg,Fe)_{5-6}(Al)_{1-2}(Si,Al)_8O_{22}(OH,F)_2$ (hydrous magnesium-iron aluminum silicate), differs from anthophyllite in that Al is substituted for some Mg, Fe, and Si. It is formed in similar environments, but with more Al present. Members of anthophyllite-gedrite series are relatively common as fine fibrous material used in commercial asbestos. This material is derived from metamorphic alteration of ultrabasic (silica-poor) rocks. Gedrite has essentially the same structure and properties as anthophyllite. Found chiefly in Montana, Idaho, North Carolina, U.S.S.R., Japan, Scotland, Finland, and India.

PARGASITE, $NaCa_2Mg_4(Al,Fe)Si_6Al_2O_{22}(OH,F)_2$ (hydrous sodium calcium magnesium aluminum-iron silicate), is included here because it illustrates the use of names to identify chemical varieties of a complex mineral. Pargasite is a variety of hornblende that contains about twice as much aluminum and sodium as the common variety. Giving this variety its own name indicates these facts, helping the geologist interpret the history of the rocks in which the variety is found. Crystal structure and properties are essentially the same as for common hornblende.

GLAUCOPHANE, $Na_2Mg_3Al_2Si_8O_{22}(OH,F)_2$ (hydrous sodium magnesium aluminum silicate), represents the alkali amphiboles, which contain more Na and less Ca than hornblendes. They are formed in metamorphic areas by alteration of Na-rich rocks or by introduction of Na in solution into the area of metamorphism. Glaucophane is the Mg-rich end member of a solid-solution series with riebeckite, the Fe-rich end member, which is found in igneous rocks. Properties of glaucophane-riebeckite minerals are like those of other amphiboles.

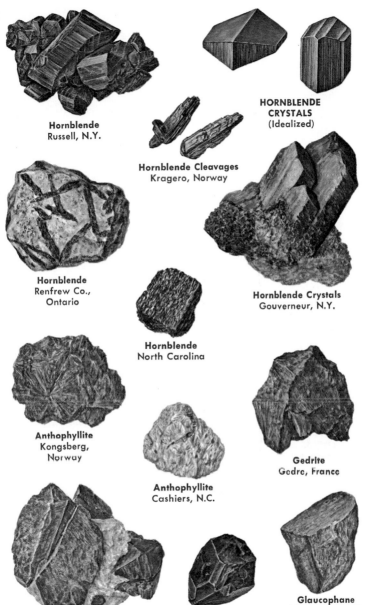

Hornblende
Russell, N.Y.

HORNBLENDE CRYSTALS
(Idealized)

Hornblende Cleavages
Kragero, Norway

Hornblende
Renfrew Co.,
Ontario

Hornblende Crystals
Gouverneur, N.Y.

Hornblende
North Carolina

Anthophyllite
Kongsberg,
Norway

Anthophyllite
Cashiers, N.C.

Gedrite
Gedre, France

Pargasite
Warwick, N.Y.

Pargasite Crystal
Pargas, Finland

Glaucophane
Marin Co., Calif.

SHEET SILICATES

In sheet silicates, each silica tetrahedron, SiO_4, shares 3 oxygen ions, O^{-2}, with adjacent tetrahedra (illustration, p. 159). The result is a sheet extending indefinitely in two directions. The charges on the shared oxygens are neutralized, and the unshared oxygen on the apex carries a -1 net charge. As in double-chain silicates, each of the hexagonal holes among the tetrahedra of a sheet must be occupied by a hydroxyl ion, OH^{-1}, to ensure stability. Stacks of these sheets are held together by positive ions, which satisfy the negative charges on the apices of the tetrahedra in the sheets. Due to the sharing of oxygen ions, the formula of a sheet silicate contains multiples of $(Si_2O_5)^{-2}$ units.

The chemical identity of the positive ions between sheets and the way in which they are placed determine the composition and structure of a silicate. A wide range of ionic sizes and charges can be accommodated. Consequently, sheet silicates, like double-chain silicates, are numerous, are complex, and show wide ranges of solid solution.

There are two important types of sheet structure: two-layer and three-layer. A two-layer sheet consists of a tetrahedral layer, with hydroxyl ions, and an octahedral layer made up of aluminum (Al) or magnesium (Mg) ions and hydroxyl ions. The positive Al or Mg ions satisfy the negative charges on the tetrahedral layer. The additional hydroxyl ions are needed to shield the positive ions and stabilize the structure. All charges are neutralized within the sheet, and hydrogen bonding holds sheets together weakly. The kaolinite and serpentine minerals have this type of sheet structure.

Three-layer sheets consist of two silica tetrahedral layers, with apices pointed toward each other. The charges on the layers are satisfied by positive ions—Mg, Al, and Fe (iron), most commonly. The oxygens on the apices of the tetrahedral layers and the hydroxyl ions in the hexagonal holes shield the positive ions and form an octahedral layer. All charges are neutralized within the sheets, and van der Waals forces hold sheets together. Talc and pyrophyllite have this structure.

Because aluminum ions, Al^{+3}, can be substituted for silica ions, Si^{+4}, a charge deficiency may exist. This may be satisfied by large ions—potassium, K^{+1}; sodium, Na^{+1}; and calcium, Ca^{+2}—located between the three-layer sheets. This provides electrostatic attraction between sheets. The micas and montmorillonites have this structure. Chlorite is somewhat similar.

Because the bonding is weak between sheets, these silicates exhibit perfect one-directional cleavage. The weak hydrogen or van der Waals bonding in kaolinite, serpentine, talc, and pyrophyllite allows easy slippage between sheets. These materials have a greasy feel and lubricating properties. The bonds between the sheets and the interlayer cations in micas are also weak, so the sheets separate easily.

COMPOSITIONS OF THE SHEET SILICATES

TWO LAYER SILICATES

Serpentine $Mg_3Si_2O_5$ $(OH)_4$

Septechlorites $(Mg, Al, Fe)_3$ $(Si, Fe)_5$ $(OH)_4$
Includes Amesite, Greenalite, Chamosite, and Cronstedtite.

Kaolinite $Al_2Si_2O_5$ $(OH)_4$
Includes Nacrite, Dickite, and Halloysite

THREE LAYER SILICATES

Talc Group—No ions between layers
    Talc $Mg_3Si_4O_{10}$ $(OH)_2$

    Pyrophyllite $Al_2Si_4O_{10}$ $(OH)_2$

Mica Group—K^+ ions between layers
Muscovite	$K\,Al_2$ $(Si_3Al)O_{10}$ $(OH)_2$
Sericite	Fine grained or "shredded" Muscovite
Illite	Clay mineral similar to Muscovite
Annite	$K\,Fe_3$ $(Si_3Al)O_{10}$ $(OH)_2$
Biotite	$K(Mg, Fe)_3$ $(Si_3Al)O_{10}$ $(OH)_2$ Mg:Fe less than 2:1
Phlogopite	$K(Mg, Fe)_3$ $(Si_3Al)O_{10}$ $(OH)_2$ Mg:Fe more than 2:1
Glauconite	Clay mineral with complex formula similar to Biotite
Zinnwaldite	$K(Li, Fe, Al)_3(Si, Al)_4O_{10}$ $(OH)_2$
Lepidolite	$K(Li, Al)_3(Si, Al)_4O_{10}$ $OH)_2$

Brittle Micas—Ca^{+2} and Na^{+1} between layers
Paragonite	$NaAl_2(Si_3Al)O_{10}$ $(OH)_2$
Margarite	$CaAl_2(Si_2Al_2)O_{10}$ $(OH)_2$
Prehnite	$Ca_2Al(Si_3Al)O_{10}$ $(OH)_2$
Clintonite	$Ca(Mg, Al)_3(Si, Al)_4O_{10}$ $(OH)_2$
Xanthophyllite	Similar to Clintonite

Montmorillonites—A group of clay minerals with a formula approximating (Ca, Na) $(Al, Mg, Fe)_4$ $(Si, Al)_8O_2$ $(OH)_4 \cdot nH_2O$

Chlorites $(Mg, Al, Fe)_6(Si, Al)_4O_{10}$ $(OH)_8$
Includes Penninite, Chlinochlore, and others

Vermiculite $(Mg, Ca)(Mg, Fe, Al)_6$ $(Al, Si)_8O_{20}$ $(OH)_4$

Sepiolite and Attapulgite—Lathlike structures
similar to sheet structures

KAOLINITE GROUP, also called the kandite group, includes kaolinite and three varieties with the same composition but a different structure. The structural unit is a two-layer sheet consisting of a silica tetrahedral layer (with associated hydroxyl ions) and an aluminum octahedral layer. The aluminum ions are coordinated with the oxygen and hydroxyl ions of the tetrahedral layer on one side and with hydroxyl ions on the other. Because two-thirds of the octahedral positions are filled, the kaolinite minerals are referred to as dioctahedral. Bonding between adjacent sheets is very weak; hydrogen bonds are present between the hydroxyl ions on one sheet and the oxygen ions of the adjacent sheet. Kaolinite minerals lose water at elevated temperatures and become refractory.

KAOLINITE, $Al_2Si_2O_5(OH)_4$ (hydrous aluminum silicate), is an important raw material used in pharmaceuticals and in firebrick, china and whitewares and a whitening agent. It occurs most commonly as a product of chemical weathering of feldspars; sedimentary processes transport, sort, and deposit it in large, pure beds from which it may be removed and used for many purposes without refinement. Kaolinite also occurs as a hydrothermal alteration product of silicates in and around sulfide veins and hot springs and geysers. It occurs generally in a very fine-grained form as clay, but some large single crystals have been found in hydrothermal deposits.

Kaolinite crystallizes in triclinic system as tabular crystals, wormlike aggregates, claylike masses, and particles scattered through sedimentary rocks. It is commonly plastic when wet, has a greasy feel, and is white if pure, colored if not. Sheets are transparent, aggregates opaque. Cleavage is perfect in one direction, hardness 2-2.5, and specific gravity 2.6. It occurs in extensive layers under cool seams or as residual weathering product in France, England, West Germany, China, Pennsylvania, Virginia, South Carolina, Georgia, and Illinois.

DICKITE also has composition of kaolinite, with layers stacked to provide monoclinic symmetry (different from nacrite). It is hydrothermal in origin; why it is formed instead of nacrite or kaolinite is not fully understood. Not distinguishable from Kaolinite without X-rays.

NACRITE is essentially the same as kaolinite, but layers are stacked differently, causing symmetry to be monoclinic. It seems to be exclusively hydrothermal in origin, is found with quartz and hydrothermal sulfides.

HALLOYSITE has the kaolinite composition, generally with excess water, probably between layers. Electron microscopic studies show that some halloysite occurs as tubular crystals (as does serpentine), suggesting that it is a highly weathered kaolinite with layers rolled up. Properties are like kaolinite's, except for water content. Most specimens show little plasticity when wet. Formation of halloysite is not well understood. It is found in small quantities in many clay deposits, particularly in Mexico, and in a sizable body at Bedford, Ind.

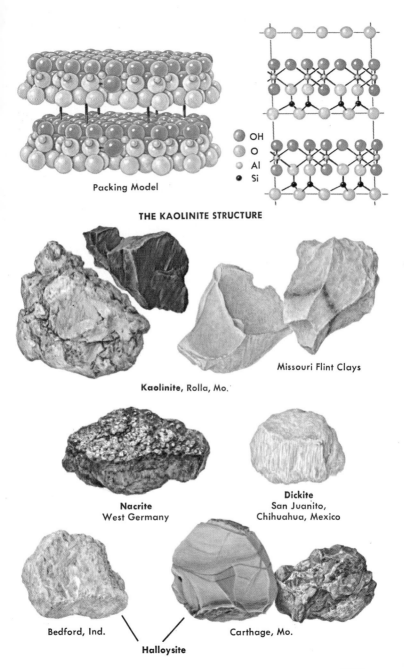

Packing Model

OH
O
Al
Si

THE KAOLINITE STRUCTURE

Missouri Flint Clays

Kaolinite, Rolla, Mo.

Nacrite
West Germany

Dickite
San Juanito,
Chihuahua, Mexico

Bedford, Ind.

Carthage, Mo.

Halloysite

189

SERPENTINE GROUP is so named because the minerals have a mottled green pattern like that of a snake. They are chemically simple, with a composition of almost exactly $Mg_3Si_2O_5(OH)_4$ (hydrous magnesium silicate), but are structurally very complex. All are very fine-grained. Because they have many different habits, the serpentines have been given many names, but only three are generally recognized: antigorite, lizardite, and chrysotile. These are not given separate treatment here because of the difficulty of distinguishing one from the other without X-ray and microscopic techniques.

The structural relationship among the serpentines has not been fully determined. All have a kaolinite-type structure, but with magnesium ions in the octahedral positions rather than aluminum ions. Since all octahedral positions are filled, the group is called trioctahedral.

Antigorite appears to have undulations in its structural layers, perhaps with some segments upside-down in relation to others. It normally occurs as masses of very small crystals, but a fibrous variety, called picrolite, has been found. Lizardite occurs as a fine-grained platy material, commonly mixed with chrysotile. Chrysotile is fibrous. It occurs in veins, with the fibers (up to 6 in. long) aligned across the veins. Evidence suggests that the fibers are tubular, formed from rolled-up layers.

The reasons why the serpentines differ in form are not known. Minor impurities, variations in pressure and temperature of formation, differences in water content, or the kinds of minerals present during formation—any or all may be involved. Serpentines are formed below 500 degrees C., in the presence of water vapor, from metamorphic alteration of ultrabasic rocks containing olivines, amphiboles, and pyroxenes. Possibly they may also be formed directly from an intruded magma of serpentine composition.

The fibrous varieties of serpentine, like those of the amphiboles, are valuable for manufacture of fire-resistant asbestos cloth. Chrysotile fibers are ideal for weaving, but are less resistant to acids than are amphibole fibers. At high temperatures serpentines lose water and recrystallize to a mixture of olivine and enstatite. This mixture is useful in making firebrick.

Recent medical studies have shown that very fine asbestos fibers can be carcinogenic, causing lung cancer in some persons who inhale the fibers over long periods of time.

Serpentines crystallize in monoclinic or orthorhombic system as platy or fibrous masses. They are green, white, yellow, gray, or greenish blue; have silky, pearly, or dull luster; are opaque to translucent; have white, commonly shiny, streak. They are decomposed by hydrochloric and sulfuric acids. Platy varieties have perfect cleavage in one direction; fibrous varieties split easily parallel to fibers. Hardness is 2.5-3, specific gravity 2.5-2.6. Widespread in New England, Texas, Maryland, Arizona, New Jersey, Pennsylvania, Quebec, E. & W. Germany, Austria, Italy, Norway, England, South Africa, and Australia.

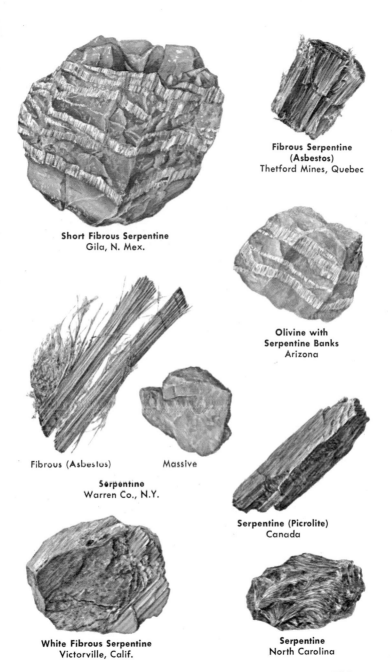

Short Fibrous Serpentine
Gila, N. Mex.

**Fibrous Serpentine
(Asbestos)**
Thetford Mines, Quebec

**Olivine with
Serpentine Banks**
Arizona

Fibrous (Asbestos)   Massive
Serpentine
Warren Co., N.Y.

Serpentine (Picrolite)
Canada

White Fibrous Serpentine
Victorville, Calif.

Serpentine
North Carolina

191

TALC, $Mg_3Si_4O_{10}(OH)_2$ (hydrous magnesium silicate), has a three-layer sheet structure (p. 186). Two silica tetrahedral layers enclose an octahedral layer in which all octahedral positions are filled with Mg ions (trioctahedral). Because the bonds between sheets are weaker than in kaolinite, talc is a better lubricant and feels greasier. It is formed by hydrothermal alteration of basic rocks and by low-grade metamorphism of silica-rich dolomite. Intergrowths of talc and tremolite have been observed, illustrating the alteration of tremolite to talc. The calcium impurity observed in commercial talcs is normally from tremolite. Talc is used as the base for talcum powder, as a ceramic raw material, and as an insulating material. Steatite and soapstone are massive impure talcs which are cut and used as acid-resistant sink and counter tops in chemical laboratories.

 Talc crystallizes in monoclinic system as tabular crystals (rare) and as foliated, radiating, and compact masses. It is white, greenish, bluish, or brownish, with white streak; is translucent to transparent; has pearly luster in large sheets; and feels greasy. Sheets are flexible, but not elastic as in micas. Cleavage is perfect in one direction, fracture irregular. Hardness is 1, specic gravity 2.58-2.83. Talc contains some iron, calcium, and aluminum. Found in many metamorphic areas—New England to Quebec, North Carolina, England, Austria, W. Germany, France, India, and China.

PYROPHYLLITE, $Al_2Si_4O_{10}(OH)_2$ (hydrous aluminum silicate), is essentially identical to talc in structure except that only two-thirds of the octahedral positions are filled. It is therefore almost identical to talc in physical properties. A metamorphic mineral derived from alteration of granite rocks, it is commonly found associated with silica and with alumina minerals such as kyanite. At high temperatures it loses water and recrystallizes as mullite and silica. It has been used as an electrical insulator and as a ceramic raw material, but is less abundant than kaolinite or talc, both similarly used. It is interesting primarily because of its structural relationship with the micas (p. 196). When K^{+1} ions are present in the environment of formation, they can be accommodated between the layers, forming micas rather than pyrophyllite. Consequently, pyrophyllite is found in rocks deficient in K^{+1} or as an alteration product of rocks from which K^{+1} has been removed.

Pyrophyllite crystallizes in monoclinic system as foliated, radiating aggregates and granular masses. It is white, greenish, yellow, or gray, with white, shiny streak; pearly or dull (massive varieties) in luster; opaque to transparent. Has perfect cleavage in one direction, uneven fracture. Contains some titanium, iron, magnesium, calcium, sodium, and potassium. Hardness is 1-2, specific gravity 2.65-2.90. Found in metamorphic areas in Carolinas, Georgia, California, Pennsylvania, U.S.S.R., Switzerland, Belgium, Sweden, and Brazil.

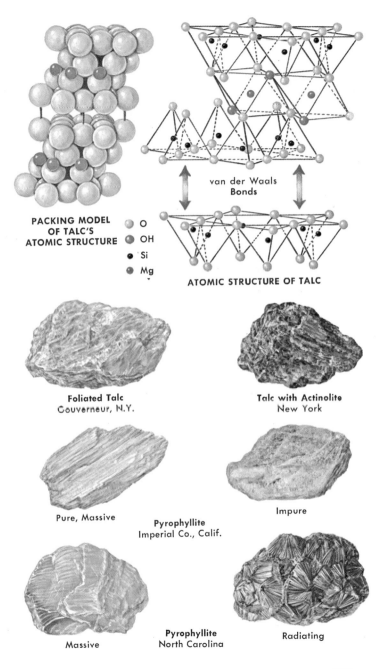

PACKING MODEL
OF TALC'S
ATOMIC STRUCTURE

- O
- OH
- Si
- Mg

van der Waals
Bonds

ATOMIC STRUCTURE OF TALC

Foliated Talc
Gouverneur, N.Y.

Talc with Actinolite
New York

Pure, Massive

Pyrophyllite
Imperial Co., Calif.

Impure

Pyrophyllite
North Carolina

Massive

Radiating

MICA GROUP minerals are derived from the ideal structure of talc or pyrophyllite in environments containing the ions of aluminum, Al^{+3}, and potassium, K^{+1}, sodium, Na^{+1}, or calcium, Ca^{+2}. The Al^{+3} replace some silica ions, Si^{+4}, in the tetrahedral positions of a sheet. Each replacement gives the electrically neutral sheet a net negative charge, but such charges are balanced by the addition between sheets of K^{+1}, Na^{+1}, or Ca^{+2}. Because aluminum, potassium, sodium, and calcium are common elements, the micas are much more abundant than either talc or pyrophyllite.

MUSCOVITE, $KAl_2(Si_3Al)O_{10}(OH)_2$ (hydrous potassium aluminum silicate), is the commonest mica. Only the feldspars and quartz are more abundant in the crust. Its structure is derived from that of pyrophyllite by replacement of one-fourth of the tetrahedral Si^{+4} by Al^{+3}, with K^{+1} added between the sheets to maintain electrical neutrality. Bonding between the sheets is much weaker than across them, as shown by the perfect cleavage in one direction, but is much stronger than in pyrophyllite, which accounts for the absence of the greasy feel and lubricating properties of muscovite. It is formed as a primary igneous mineral in granites and other silica-rich rocks and in a wide range of metamorphic rocks. It is also a major constituent of shales.

Muscovite crystallizes in monoclinic system as tabular crystals, often pseudohexagonal; is dispersed in igneous rocks as books, in schists and gneisses as bands, and in sedimentary rocks as clay. It has perfect one-directional cleavage, splitting into sheets that tear with hackly edges, but are elastic and strong. It is colorless, gray, pale green, brown, yellow, pink, or violet, with white streak; glassy or pearly in luster; opaque to translucent, transparent in thin sheets. Hardness is 2.5-3, specific gravity 2.7-3.0. Found most everywhere.

BIOTITE, $K(Fe,Mg)_2(Si_3Al)O_{10}(OH)_2$ (potassium magnesium-iron aluminum silicate), is derived from the talc structure by substitution of Al^{+3} for Si^{+4} and inclusion of K^{+1} between sheets. Much iron, as Fe^{+2}, is present in octahedral positions. Composition is more varied than in muscovite. Biotite occurs in igneous rocks from granites to gabbros and many lavas, and in a wide variety of metamorphic rocks. A number of biotite minerals containing other elements have received mineral names. Among these are haughtonite and siderophyllite, with abundant iron; manganophyllite, with manganese; wodanite and titanobiotite, with titanium; and calciobiotite, with abundant calcium.

Biotite crystallizes in monoclinic system as tabular prismatic crystals, scattered grains and scales, and scaly masses. It cleaves in one direction, splitting into elastic sheets. It is black, dark green, or brown, with white streak; glassy in luster; opaque to transparent. It is bleached by sulfuric acid. It occurs throughout world.

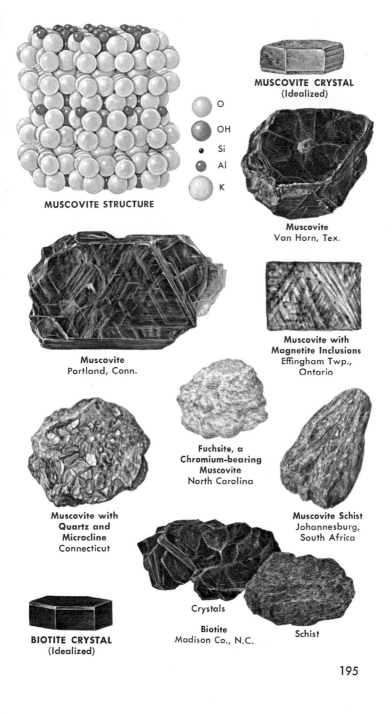

O
OH
Si
Al
K

MUSCOVITE STRUCTURE

MUSCOVITE CRYSTAL
(Idealized)

Muscovite
Van Horn, Tex.

Muscovite
Portland, Conn.

Muscovite with Magnetite Inclusions
Effingham Twp., Ontario

Fuchsite, a Chromium-bearing Muscovite
North Carolina

Muscovite with Quartz and Microcline
Connecticut

Muscovite Schist
Johannesburg, South Africa

BIOTITE CRYSTAL
(Idealized)

Crystals

Biotite
Madison Co., N.C.

Schist

PHLOGOPITE, $K(Mg,Fe)_3(Si_3Al)O_{10}(OH)_2$ (hydrous potassium magnesium-iron aluminum silicate), is essentially a biotite (p. 194) with Mg more than twice Fe. It occurs notably in metamorphosed siliceous dolomites and some ultra-basic rocks. It is common at contacts between igneous intrusives and dolomite country rocks and as an accessory mineral in dolomite marbles. It commonly shows asterism (starlike figure) when viewed in transmitted light.

PARAGONITE, $NaAl_2(Si_3Al)O_{10}(OH)_2$ (hydrous sodium aluminum silicate), differs from muscovite (p. 194) in having Na^{+1} instead of K^{+1} between sheets. Found in metamorphic areas, mostly in banded rocks; in some quartz veins; and in some sediments. It occurs as massive scaly aggregates (monoclinic); has perfect one-directional cleavage; is pale yellow, gray, or green, with pearly luster. Hardness is 2.5, specific gravity 2.85. Occurs notably in the Alps of Europe.

Phlogopite crystallizes in monoclinic system as prismatic crystals and as flakes and sheets, disseminated or in oriented bands. Color is yellow, brown, green, white; luster commonly pearly with goldlike reflection. Sheets are elastic and strong, but tear; are transparent if thin. Cleavage is perfect in one direction; hardness 2-2.5; specific gravity 2.76-2.90. Found chiefly in New England, Quebec, Italy and Switzerland (Alps), Malagasy Rep., Ceylon, Finland, and Sweden.

MARGARITE, $CaAl_2(Si_2Al_2)O_{10}(OH)_2$ (hydrous calcium aluminum silicate), is commonest "brittle mica." It is harder than muscovite, its sheets more brittle. It occurs with corundum as laminar aggregates and scaly masses (monoclinic system) in metamorphic areas of Appalachians, European Alps, Urals, Greece, Turkey. It is gray, white, pink, or yellow; has perfect one-directional cleavage. Hardness 3.5-4.5, specific gravity 3-3.1.

PREHNITE, $Ca_2Al(AlSi_3)O_{10}(OH)_2$ (hydrous calcium aluminum silicate), is found in cavities and veins of basic volcanic rocks (associated with zeolites), in impure metamorphic limestones, and in some veins of igneous rocks. It crystallizes in orthorhombic system as barrel-shaped crystals, small radiating clusters, globular masses. It is brittle (a brittle mica) and has good cleavage in one direction. Color—shades of green, white, gray—commonly fades on exposure to air. Hardness 6-6.5, specific gravity 2.90-2.95. Found in Appalachians (Mass., Conn., N.J.), Michigan, European Alps, Scotland, South Africa, and elsewhere.

LEPIDOLITE, $K_2(Li,Al)_{5-6}(Si_{6-7}Al_{2-1})O_{20}(OH,F)_4$ (hydrous potassium lithium-aluminum silicate), has muscovite structure (p. 196) with Li^{+1} replacing some Al^{+3}. A popular collector's item because of pink or purple colors, it is commonest source of lithium. Added to glasses and enamels, it lowers thermal expansion, thus improves heat resistance. It is found exclusively in pegmatites as large flakes or disseminated grains, notably in Maine, Connecticut, North Carolina, California, U.S.S.R., Czechoslovakia, E. Germany, Italy, Malagasy Rep. It is commonly associated with two other lithium minerals, tourmaline and spodumene and with K-feldspar and quartz.

GLAUCONITE has a complex mica-type formula, a potassium iron silicate. Composition varies greatly, but name is given to greenish mineral that forms generally in marine environment. It is a major constituent of marine "greensands." Found in many localities, notably Bonneterre, Mo.; Otago, New Zealand; and Ukraine; also found in modern sediments along the continental shelves.

Phlogopite, Godfrey, Ontario

with Tourmaline

Lepidolite
Pala, Calif.

**Paragonite Schist, with
Kyanite and Staurolite**
Canton Ticino, Switzerland

Margarite
Chester, Mass.

Lepidolite
Coolgardie, Australia

Prehnite
Duluth, Minn.

Prehnite
Upper Montclair,
N.J.

Glauconite
Llano, Tex.

197

CLAY MINERALS

The word clay is generally applied to any rock or soil material having a significant percentage of very small particles and exhibiting plasticity (the ability to be molded when wet). Particle size is on the order of 2 microns or less in diameter (1 micron = 0.001 millimeter, or 0.00004 inch). Some clays, when fired, develop desirable properties—hardness, impermeability, pleasing colors—and are valuable ceramic materials. Most soils contain some clay particles. These largely determine the soil's suitability for cultivation and the engineering requirements for construction of roads, buildings, or other structures on the soil. The study of clays is still in an early stage.

STRUCTURE AND CHEMISTRY OF CLAYS were little understood until the middle of this century primarily because of the small particle sizes. Then, the development of X-ray diffraction for structural analysis and of the electron microscope for direct observation of small particles clearly established that the unique properties of clays are the direct result of the sizes and structures of the individual particles.

Clays are, in general, weathering products of pre-existing rock. They may have remained where the weathering occurred or may have been transported by water, wind, or ice to other areas, having been sorted and possibly concentrated in the process. Resistant minerals, such as quartz; incompletely weathered minerals, such as feldspar; and minerals formed by precipitation from aqueous solution, notably calcite, may be present. But it is the presence of the clay minerals, formed by complete chemical transformation of other silicates, that determines the characteristics of clay.

Clay minerals are hydrolyzed silicates (contain hydroxyl ions) or aluminosilicates of the sheet type. The major ones are described on pages 200-203 and are classified, with a summary of uses, on the facing page. Clays are particularly useful as aids to geologic understanding and as raw materials for industrial products of many kinds.

The chemical composition of a clay depends not only on the mineral from which it was formed, but also on the chemical environment of weathering, which in turn is largely the result of climate. Thus, in basic soils of warm-climate regions, weathering tends to remove alkalies and silica and leave aluminum and iron; the result is kaolinite formation and typical lateritic, or red, soils. In acid soils of temperate climates, weathering tends to remove all but the silica, lowering the clay mineral content. If incomplete weathering is the rule, as in arid climates, montmorillonites and chlorites may form and persist. Study of the compositions and structures of clay minerals in rocks and soils therefore may provide valuable evidence in interpreting the chemical and climatic conditions under which the rocks and soils were formed.

COMMERCIAL USES OF CLAYS depend on several properties. The sheet-silicate structures and small particle sizes of clays, with the resultant large surface areas per unit weight, are responsible for their affinity for water, their plasticity, and their ability to absorb some organic molecules. Such properties make some clays, notably the montmorillonites, useful as Fullers earth for cleaning and filtering processes, as fillers for plastics and rubber, and as an agent for suspending particles in water (e.g. in ceramic slips and drilling muds).

When heated, clays lose the water of hydrolysis and become strong and more or less impermeable. These properties make clays useful in the manufacture of bricks, tile, and pottery. Kaolinite is able to withstand high temperatures without deforming and is therefore important for making firebrick and other refractory materials. These are a few of the thousands of applications for which clays are suited as a result of their unique properties.

CLAY MINERAL CLASSIFICATION AND USE

Structural Type	Mineral	Uses
2-Layer	Kaolinite Nacrite Dickite Halloysite	As raw material for firebrick, china, whiteware, porcelain, tiles; as paper filler.
3-Layer Non-expandable	Illite (Mica)	Some use as raw material for architectural brick, sewer pipe; bloating varieties for ornamental clay; for molding sands.
	Glauconite	Not sufficiently abundant or concentrated to be useful.
	Chlorite	Not sufficiently abundant or concentrated to be useful.
3-Layer Expandable	Montmorillonite	As fillers for many materials, as filters, as fuller's earth, as a suspending agent.
	Vermiculite	Because it expands when fired, used as mulching agent, thermal insulation, and packing material.
Fibrous	Sepiolite	Carved as found (Meerschaum) into pipes and ornaments.
	Attapulgite	As suspending agent for ceramic slips.

MONTMORILLONITE, $(Ca,Na)_{0.35-0.7}(Al,Mg,Fe)_2$ $(Si,Al)_4$ $O_{10}(OH)_2 \cdot nH_2O$ (hydrous calcium-sodium aluminum-magnesium-iron silicate), is a general name for a group of clay minerals with widely varying compositions. They have a structure essentially like that of the micas, but differ in important respects. Water can penetrate their structure and cause the clay to expand. On heating or prolonged exposure to dry air, the water may leave the structure, allowing it to collapse. These clays are sensitive, therefore, to changes in moisture content and present severe engineering problems when present in soils. Montmorillonites form by weathering or hydrothermal alteration of basic igneous rocks, notably volcanic ash. They are common constituents of soil materials in many areas. Of the many uses for montmorillonite, one of the more ingenious is in carbonless paper. Because some organic chemicals cause montmorillonite to turn blue, the lower sheet is sprayed with a thin layer of montmorillonite; the bottom of the upper sheet is sprayed with the liquid, encapsulated in microscopic gelatin bubbles. When written or typed upon, the bubbles break, releasing the liquid and turning the montmorillonite blue, producing a clean copy on the lower sheet.

Montmorillonites crystallize in the monoclinic system, occurring as very small particles packed in massive layers or mixed with other minerals in soil. They are soft and plastic when wet, brittle when dry. Color is white, yellowish, greenish, or gray. Cleavage is perfect in one direction. Hardness is 1-2, specific gravity 2-3. They are sometimes called smectites.

Varieties include beidellite (Al-rich), nontronite (Fe-rich), saponite (Mg-rich), hectorite (Li-bearing), and sauconite (Zn-bearing). Large deposits occur in Wyoming (as bentonite layers) and in France, E. Germany, British Isles, and many other areas. The absorbing properties of these minerals make them useful in industry.

CHLORITE, $(Mg,Fe,Al)_6(Si,Al)_4O_{10}(OH)_8$ (hydrous magnesium-iron-aluminum silicate), is similar to the micas in that it consists of three-layer talc-like units. But instead of K^{+1} ions between layers, it has a brucite-like layer with some Al^{+3} or Fe^{+3}. Chlorites of varying composition are major constituents of low-grade metamorphic rocks called greenschists. They also occur as weathering or hydrothermal alteration products of basic silicates such as pyroxenes, amphiboles, and biotites.

Chlorite crystallizes in monoclinic system as tabular masses and disseminated scales. It is most commonly green, but may be brown, yellow, white, pink, or red; or may weather, with oxidation of iron, to rusty color. It has perfect cleavage in one direction and pearly luster on cleavage surfaces. Transparent in thin sheets, it is not as elastic as micas and is easily attacked by strong acids. Hardness is 2-3, specific gravity 2.6-3.3. It is found in many areas of regional metamorphism as a late-stage alteration of many igneous rocks, and is also found in numerous sediments. There are many varieties of the mineral, notably penninite, chlinochlore, and prochlorite.

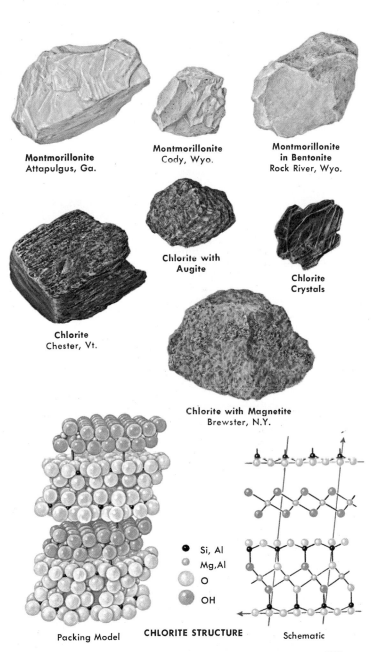

Montmorillonite
Attapulgus, Ga.

Montmorillonite
Cody, Wyo.

**Montmorillonite
in Bentonite**
Rock River, Wyo.

**Chlorite with
Augite**

**Chlorite
Crystals**

Chlorite
Chester, Vt.

Chlorite with Magnetite
Brewster, N.Y.

● Si, Al
○ Mg, Al
○ O
○ OH

Packing Model   **CHLORITE STRUCTURE**   Schematic

SEPIOLITE, $Mg_9Si_{12}O_{30}(OH)_6 \cdot 10H_2O$ (hydrous magnesium silicate), is not strictly a sheet silicate, but is similar in structure to talc. It has lath-like structural units, with channels extending through the crystal in one direction. In the mineral as it occurs in rocks, the channels are filled with water molecules. The water can be driven out by heating, making the crystal very reactive and highly absorbent. Meerschaum, used for pipe-making, is a lightweight aggregate of sepiolite with other minerals.

Sepiolite is a white powdery material believed to crystallize in monoclinic system (electron microscope shows fibrous crystals). It occurs in granites and syenites and in magnesium-rich rocks; also as sedimentary mineral associated with brines, particularly magnesium-rich playa deposits. Hardness 2-2.5, specific gravity about 2. Used as fuller's earth. Found in New Mexico, Pennsylvania, Utah, Nevada, Turkey, Greece, Morocco, U.S.S.R., Shetland Island, Arabia, Spain.

ATTAPULGITE, $Mg_5Si_8O_{20}(OH)_2 \cdot 8H_2O$ (hydrous magnesium silicate), is structurally similar to sepiolite, but with narrower laths. Chemical and environmental relationships between the two are not completely understood. Both appear to be formed by alteration of montmorillonite in magnesium-rich waters, attapulgite being the less thoroughly altered stage. It generally contains more aluminum in solid solution, which may be the determining factor in its structure.

Attapulgite crystallizes in monoclinic system as white claylike material (fibrous under electron microscope). Other data are sparse. Large deposits at Attapulgus, Ga., are mined for use in ceramic industry. It is widespread as sedimentary mineral associated with sepiolite. Not enough is known to identify and separate the two in all cases.

MOLECULAR SIEVES are interesting materials whose properties depend on the existence of large channels through the structure, as in sepiolite (see facing page). It has long been known that small atoms, such as hydrogen and helium, have the ability to permeate and pass through solid materials. Only recently, however, has it been shown that crystals such as sepiolite, attapulgite, and some of the zeolites (pp. 226-227) can act as filters, allowing small molecules to pass through the channels of the structure rather rapidly while blocking larger molecules. These materials have found uses in purification of liquids.

FULLER'S EARTHS have been used for centuries to remove oils and dirt from wool; the fuller mixed small quantities of certain clays with water for washing purposes. Now it is realized that the cleaning occurs because some clays, such as montmorillonites (pp. 200-201), being extremely fine grained, can absorb some molecules onto the crystal surface. A similar absorption is responsible for the ability of some montmorillonites to remove objectionable molecules from wines and beer.

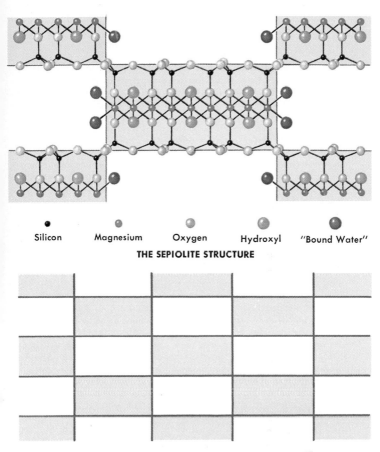

| Silicon | Magnesium | Oxygen | Hydroxyl | "Bound Water" |

THE SEPIOLITE STRUCTURE

**DIAGRAMMATIC REPRESENTATION OF SEVERAL UNIT
CELLS OF SEPIOLITE SHOWING CHANNELS THROUGH STRUCTURE**

**Sepiolite
(Meerschaum)**
Media, Pa.

Meerschaum
Pipe

Attapulgite
Attapulgus, Ga.

FRAMEWORK SILICATES

In framework silicates, unlike other silicates, the oxygen ions at all four corners of the silica tetrahedron are shared with adjacent tetrahedra. The oxygen ions are therefore completely saturated. There is no residual charge on the tetrahedron, and the structural unit is infinite in three dimensions. The prototypes of the framework silicates are the silica minerals having the formula SiO_2. These minerals are quartz, tridymite and cristobalite.

The framework structure has two characteristics that are important to note: (1) the structure is necessarily "open," with large interstitial positions, and (2) the number of possible arrangements is large. Also important is that the aluminum ion, Al^{+3}, is stable in the tetrahedral positions. An oxygen shared between two tetrahedra, one of which is occupied by Al^{+3} instead of Si^{+4}, is not saturated, but has a net negative charge. This must be balanced by another positive ion in an adjacent interstitial position. A general formula may be written: $M_x^{+1}Al_xSi_{1-x}O_2$, where M represents ions in interstitial positions, and $Al_xSi_{1-x}O_2$ shows the tetrahedral makeup. Because the interstitial positions are large, relatively large ions—e.g. Na^{+1}, K^{+1}, Rb^{+1}, Cs^{+1}, Ca^{+2}, Sr^{+2}, Ba^{+2}, or Pb^{+2}—are stable in such a structure.

Given the possible variations in framework configuration and the number of common large ions, the number of framework silicates is large. The feldspars (sodium, potassium, and calcium aluminosilicates) and quartz are common framework silicates. Their abundance in the earth is great because the elements that comprise them are plentiful and their framework structure is chemically and thermally stable.

Because of the ease with which the framework can be modified to accommodate ions of different sizes without changing the bonding scheme, there is extensive solid solution (p. 32) among framework silicates. The plagioclase feldspars, for example, can have calcium and sodium present in any ratio. Because the framework linkage can be maintained sufficiently open to accommodate water molecules, some framework silicates, notably the zeolites, have water in the structure.

Of particular interest to the mineralogist and geochemist is the distribution of elements, particularly minor elements, among minerals of different structural types. This information is of value in tracing the crystallization sequences of minerals and in establishing relationships among different rocks. The framework silicates, particularly the feldspars, because of their ability to accommodate a variety of large metal ions, are especially interesting in this regard. Feldspars, being the most abundant mineral group in the crust of the earth, also supply the raw material for production of clay minerals by chemical weathering. For this reason, the reactions of feldspars with aqueous solutions, basic and acid, have been well studied. The information thus gained has been invaluable to the sedimentary petrologist.

FRAMEWORK SILICATES

SILICA GROUP, SiO_2.   Quartz, tridymite, cristobalite, and
a wide variety of types of quartz.

FELDSPAR GROUP
  Potassium feldspars, $K(AlSi_3)O_8$
    Microcline
    Sanidine
    Orthoclase
    Adularia

  Plagioclase feldspars
    Albite, $Na(AlSi_3)O_8$, in solid solution with
    Anorthite, $Ca(Al_2Si_2)O_8$.

  Mixed feldspars
    Perthite, $K(AlSi_3)O_8$ and $Na(AlSi_3)O_8$
    Anorthoclase, $(Na, K)AlSi_3O_8$.

  Miscellaneous feldspars
    Celsian, $Ba(Al_2Si_2)O_8$
    Strontium feldspar, $Sr(Al_2Si_2)O_8$ (synthetic)
    Lead feldspar, $Pb(Al_2Si_2)O_8$ (synthetic)

FELDSPATHOIDS*
  Nepheline, $Na_3K(Al_4Si_4)O_{16}$
  Kalsilite, $K(AlSi)O_4$
  Eucryptite, $Li(AlSi)O_4$
  Petalite, $Li(AlSi_4)O_{10}$
  Cancrinite, $(Na,Ca)_{7-8}(Al_6Si_6)O_{24}(CO_3,SO_4,Cl)_{1.5-2} \cdot 1-5H_2O$
  Vishnevite, $(Na,Ca,K)_{6-7}(Al_6Si_6)O_{24}(SO_4,CO_3,Cl)_{1-1.5} \cdot 1-5H_2O$
  Sodalite, $Na_8(Al_6Si_6)O_{24}Cl_2$
  Noselite, $Na_8(Al_6Si_6)O_{24}SO_4$ (also called nosean)
  Hauynite, $(Na,Ca)_{4-8}(Al_6Si_6)O_{24}(SO_4,S)_{1-2}$ (also called hauyne)
  Leucite, $K (AlSi_2)O_6$
  Analcite, $Na (AlSi_2)O_6 \cdot H_2O$

*A number of different silica-poor framework silicates are here
grouped under the general name feldspathoid. This is not done in
most classifications.

SCAPOLITE GROUP
  Marialite, $Na_4 (Al_3Si_9)O_{24}Cl$
  Meionite, $Ca_4 (Al_6Si_6)O_{24}CO_3$
  General Formula, $(Na, Ca, K)_4(Al, Si)_{12}O_{24}(Cl, F, OH, CO_3, SO_4)$

ZEOLITE GROUP
  General Formula, $(Na_2, K_2, Ca, Ba) (Al, Si)_nO_{2n} \cdot x H_2O$
  At least 22 different zeolite minerals have been described
  and have valid mineral names.

SILICA GROUP includes minerals with the formula SiO_2. Since silicon, Si, and oxygen, O, are the two most abundant elements in the earth's crust, silica group minerals are common everywhere. Free silica, SiO_2, occurs most commonly as quartz, either in large crystals or in very fine grains (cryptocrystalline varieties). Noncrystalline (or amorphous) silica containing water is also found in surface deposits. Two high-temperature varieties, cristobalite and tridymite, are found in surface lava flows and on the Moon. The relationships among the three common forms of crystalline SiO_2, in terms of temperature-pressure stability ranges, have been studied in detail. The rather "open" structures permit other ions to occupy interstitial positions; thus, coupled with substitution of Al^{+3} for Si^{+4} in tetrahedral positions, provides means for modification of the structures.

QUARTZ, SiO_2 (silicon dioxide), is used in the manufacture of firebrick and as a gemstone (in its many varieties), as an electronic component, and, most commonly, as the principal constituent of glass. It is one of the most thoroughly studied minerals. Its structure consists of silica tetrahedra sharing all four corners, so that each silicon ion is surrounded by four oxygen ions, and each oxygen by two silicons. Since the charges on each are satisfied, no other elements are necessary.

"Holes" in the rather open structure can accommodate smaller ions, such as lithium and zinc and some Al^{+3} can occupy tetrahedral sites. When heated, the structure opens even more, changing to a form called high-quartz at 573° C. This involves only a change in bond angles and no significant structural rearrangement. On cooling, the low-quartz structure is restored, but the large thermal expansion and contraction near 573° C may cause fracturing if the crystal is cooled very rapidly.

Nearly all surface rocks contain quartz. Exceptions are some limestones and dolomites and the basic igneous and metamorphic rocks. Single quartz crystals with well-developed faces are found in hydrothermal veins, in some sedimentary cavities, and in other rare environments where unrestricted growth can occur.

Quartz crystallizes in hexagonal system as short to long prismatic crystals, often in the form of twins. It occurs most commonly as irregular grains intergrown with other minerals in igneous and metamorphic rocks; as rounded grains in sandstones; and as microscopically grained specimens (cryptocrystalline varieties). It is colorless if pure, almost any color if not (see gem varieties, p. 212). It is transparent unless cryptocrystalline. Quartz has no streak. It is soluble in hydrofluoric acid and molten sodium carbonate. Fracture is conchoidal; good cleavage is lacking. Hardness is 7, specific gravity 2.65. In igneous rock, quartz is associated chiefly with potassium feldspars and muscovite mica. It occurs in shales as very small particles. It is widely distributed.

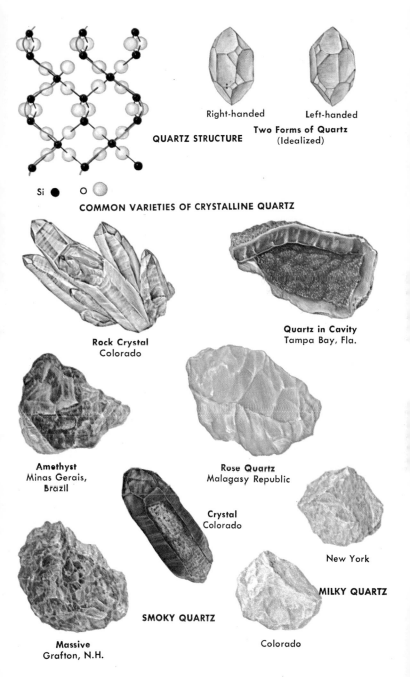

Si ●   O ◯ ◯

QUARTZ STRUCTURE

Right-handed   Left-handed

Two Forms of Quartz
(Idealized)

COMMON VARIETIES OF CRYSTALLINE QUARTZ

Rock Crystal
Colorado

Quartz in Cavity
Tampa Bay, Fla.

Amethyst
Minas Gerais,
Brazil

Rose Quartz
Malagasy Republic

Crystal
Colorado

New York

MILKY QUARTZ

SMOKY QUARTZ

Massive
Grafton, N.H.

Colorado

CRYPTOCRYSTALLINE QUARTZ is simply quartz whose crystals are so minute that they can be seen only through a microscope. It is formed from silica that has been removed from silicate minerals by chemical weathering. This silica is carried by water as ultrafine particles in colloidal suspension. It eventually settles out as a noncrystalline silica gel (amorphous silica, p. 210), containing appreciable quantities of water. The amorphous silica loses water gradually and forms very small crystals, even at low temperatures. During settling, chemical conditions are changing slowly. As they change, the color, rate of deposition, and texture of the precipitate commonly change also. The result is banding, as exhibited by the agates. Often there are several types of cryptocrystalline material in a single specimen.

Cryptocrystalline quartz occurs in many varieties. These have been given names based on their color, opacity, banding, and other observable physical features. They are found most commonly in sedimentary rocks—as fillings between sand grains, as cavity fillings in larger holes, and as bands and nodules associated with limestone and shale.

Vividly colored and banded varieties of cryptocrystalline quartz are popular as gemstones. All varieties may be generally classified as either chalcedony or chert, depending on luster and transparency.

CHALCEDONY has a waxy luster, extremely small crystals, and a specific gravity only slightly less than large quartz crystals. Nearly transparent to translucent, it will fracture in a conchoidal manner, with sharp edges. It may be massive, botryoidal, fibrous, or encrusted, and it may be associated with well-formed crystals of quartz, especially in geodes. The most common chalcedony is gray, but it may be white, brown, black, or blue. Some varieties are difficult to distinguish from opal. The only difference between chalcedony and chert is luster. Chert is dull (due to impurities); chalcedony is nearly pure SiO_2. Among the varieties are:

CHERT is duller and more nearly opaque than chalcedony. Its luster may be dull to almost glassy, and its fracture is irregular to conchoidal. Ordinary tan chert is the most abundant cryptocrystalline quartz. Chert may be white, tan, brown, black, or gray. The name jasper is generally used for the more brightly colored varieties of chert. Jasper is usually red or brown, but may also be green, blue, or yellow, or it may be banded. Flint is a black variety of chert. Aventurine is chert with sparkling scales of included mica. Because of the conchoidal fracture, chert is easily chipped so that sharp edges are formed, hence its use for cutting stones and arrowheads.

GEM VARIETIES OF CHALCEDONY

Agate—banded or with branching inclusions (moss agate). Petrified wood is commonly wood replaced by agate.

Onyx—an agate with straight, even banding. White and black bands are common.

Sardonyx—with bands of red (carnelian), white, brown, or black.

Tigereye — fibrous aggregates, commonly with wavy fibers, in a bundle. Banding across fibers.

Carnelian—clear, red-to-brown.

Chrysoprase—green chalcedony.

VARIETIES OF CHALCEDONY

Tigereye
South Africa

Blue Tigereye
South Africa

Chalcedony
San Bernadino,
Calif.

Moss Agate
New Mexico

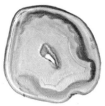

Agate
Mexico

Agate
Minas Gerais, Brazil

VARIETIES OF CHERT

Flint
Dover Cliffs, England

Chert
Joplin, Mo.

Aventurine
India

Jasper
Honduras

Jasper
Colchester, Vt.

Petrified Wood
Holbrook, Ariz.

AMORPHOUS SILICA is noncrystalline SiO_2, occurring as opal, a colloidal precipitate with abundant residual water; and as glass, formed by cooling of molten silica so rapidly that crystallization could not occur.

OPAL, a gel precipitated from an aqueous suspension, contains appreciable water. It is less dense than quartz. With time (millions of years) opal loses water and crystallizes. Opal's luster is commonly glassy or pearly, but may be resinous or dull. Pure opal is white, but impurities give it any color. Among the many varieties of opal are:

SILICA GLASS, formed by melting of silica, is scarce because of silica's high melting point (about 1710° C). Also called lechatelierite, it has been found as inclusions in volcanic rocks and as tubes (fulgurites) formed when lightning strikes quartz sandstone. At Meteor Crater, Arizona, heat generated by the meteorite's impact formed silica glass.

Common opal—white or in pale colors.

Precious opal—like common opal, but with a play of rainbow colors caused by banding and internal cracks.

Fire opal—red or yellow, with a striking play of colors.

Wood opal—petrified wood with opal as the petrifying material.

Tripolite — siliceous remains of diatoms. A fine white powder used as polishing material.

Moss opal—with branching inclusions.

Hyalite—thin crusts on rock.

GEM VARIETIES OF SILICA include crystalline quartz (p. 207), cryptocrystalline quartz (p. 210), and opals. The hardness, chemical resistance, transparency, and general attractiveness of quartz make it well suited for use as a gem. Its low index of refraction and relative abundance, however, detract from its monetary value. Gem varieties of quartz have been given many names, some misleading. Smoky topaz, for example, is not a topaz (p. 168) and is more correctly called smoky quartz. The varieties fit generally into three categories: (1) quartz crystals, either clear or colored by atomic impurities or structural disorders; (2) quartz crystals with inclusions of other minerals; and (3) quartz pseudomorphs formed by quartz replacing and assuming the shape of other minerals.

QUARTZ CRYSTALS most commonly used as gems include:

Rock crystal—colorless, transparent crystals; any size or shape.

Amethyst—purple or violet shades. Loses color when heated.

Rose quartz—pink and generally massive.

Smoky quartz—yellow to dark brown or black; also called cairngorm if yellow or brown, morion if black.

Milky quartz—white, cloudy, translucent.

Citrine—yellow quartz; also called "topaz."

SAGENITIC QUARTZ crystals contain inclusions, generally needle-like, such as hydrous iron oxides, tourmaline, rutile, chlorite, asbestos, and stibnite. Gemstones with inclusions are rutilated quartz, with thin needles of rutile; cat's eye, with asbestos fibers; and aventurine, with mica or hematite.

QUARTZ PSEUDOMORPHS of many different minerals occur. The best known gem variety is tigereye, in which quartz has replaced asbestos, but retains the fibrous structure, as shown on the opposite page.

Wood Opal
Butte, Mont.

Moss Opal
Boise, Idaho

Precious Opal
Australia

Tripolite
Santa Barbara Co.,
Calif.

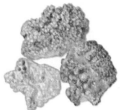

Hyalite
(Encrusted Opal)

Rutilated Quartz
Brazil

SILICA GEMSTONES
Octagon Cut

Smoky Quartz   Amethyst   Citrine   Rutilated

Cabochon Cut

Agate   Cristobalite in Obsidian   Agate   Aventurine

Mottled Jasper   Rose Quartz   Tigereye   Chalcedony

CRISTOBALITE, SiO_2 (silicon dioxide), the form of silica that is stable at high temperatures. It crystallizes in the cubic system and has a rather open structure quite different from that of quartz. At ordinary temperatures, however, the structure closes a bit, causing the crystal to become tetragonal. Pure quartz will convert to cristobalite at about 1025° C when heated without water-vapor pressure. The conversion is irreversible so that cristobalite remains on cooling.

If other materials, notably Na_2O and Al_2O_3, are present during crystallization, however, cristobalite may not be formed. Apparently the other oxides go into solid solution with SiO_2 and may cause quartz or tridymite to form over wide ranges of temperature, complicating the stability relationships.

Cristobalite was first discovered with tridymite in an andesite lava at Cerro San Cristobal, Pachuca, Mexico.

TRIDYMITE, SiO_2 (silicon dioxide), has a structure similar to that of cristobalite, different only in the arrangement of successive layers of silica tetrahedra. All natural tridymites contain oxides other than SiO_2; it has been suggested, therefore, that the structure is stabilized by those oxides and that it cannot exist as pure SiO_2. Tridymite is orthorhombic at ordinary temperatures, but the structure opens up at higher temperatures so that it is near the ideal hexagonal network shown. It has been found in silica-rich volcanic rocks in many places with other high-temperature minerals.

THE IMPORTANCE OF SILICA IN ROCKS cannot be overstated. Not only is quartz very abundant, making up about one fifth of the crust of the earth, in igneous, metamorphic, and sedimentary rocks, but silica, SiO_2, is a major component of all the common rock-forming minerals (the silicates) except calcite (as $CaCO_3$ in limestones). The feldspars, micas, amphiboles, pyroxenes, olivines, and clay minerals all contain between 40% and 70% SiO_2 in combination with the other common oxides. What minerals crystallize is determined by the amount of available SiO_2 relative to the amounts of the other oxides. The kinds of silicate minerals, in turn, determine what the weathering rates and products will be and, therefore, the kinds of soil and other sedimentary material will be formed. The silica contents of lavas determine whether volcanic activity will be violent or relatively quiet. Basic lavas, like basalt, are low in silica and crystallize to olivines, pyroxenes, etc. In the melt, the SiO_4 tetrahedra do not share many corners and are free to move. The lava has a low viscosity and thus gas pressures cannot build up. In acid lavas, like rhyolites, which crystallize to quartz, feldspars, etc., the tetrahedra are linked, the lava is viscous, and gas pressures build to explosive levels. The total silica contents of igneous and metamorphic rocks over a large area provide geologists with information for determining the history of formation of the rocks and understanding the process of planetary evolution.

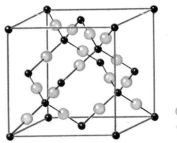

O
Si

**IDEALIZED CRISTOBALITE
STRUCTURE AT HIGH TEMPERATURE**

Cristobalite in Obsidian
Millard Co., Utah

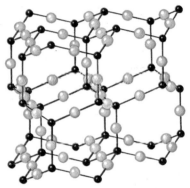

**IDEALIZED TRIDYMITE
STRUCTURE AT HIGH TEMPERATURE**

Tridymite in Lava
Eifel, West Germany

**Quartz-rich
Sandstone**

**Quartz with Microline
and Mica in Granite
Gneiss (Metamorpric)**

**Quartz Veinlets
in Igneous Rock**

**Quartzites
(Metamorphic)**

**Mottled Chert from
Sedimentary Layer**

213

THE FELDSPAR GROUP

If only the silicon ion were stable in the oxygen tetrahedron, the silica minerals would be the only framework silicates. The aluminum ion, which is also abundant, is stable in tetrahedral sites, however. If one fourth to one half the tetrahedral sites are filled by Al^{+3} ions rather than Si^{+4} ions, electrical neutrality is preserved by other positive ions in the rather large openings among the tetrahedra. Because K^{+1}, Na^{+1}, and Ca^{+2} are abundant and can fit into the interstitial positions, the feldspars—$KAlSi_3O_8$, $NaAlSi_3O_8$, $CaAl_2Si_2O_8$ and solid solutions of these—are the most abundant mineral group, making up about one half the crust of the earth.

MICROCLINE, $KAlSi_3O_8$, is the most common potassium feldspar, abundant in granites and other silica-rich igneous rocks, in many metamorphic rocks, and in sediments. Microcline differs from the other potassium feldspars in that the distribution of silica and aluminum ions in tetrahedral sites is regular (ordered) rather than random, as in sanidine, or partially ordered, as in orthoclase and adularia. Most microcline crystals do not have good faces because they have crystallized in the presence of quartz and other minerals. A single crystal is normally an intergrowth of microcline with smaller amounts of albite, $NaAlSi_3O_8$. The intergrowth, with a distinctive streaked appearance, is called perthite. Microcline is formed at high temperatures during crystallization of a magma or by metamorphism deep within the earth and the ordering occurs during the very slow cooling process.

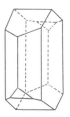

Microcline is triclinic, with cleavage in three directions near 90°. The fracture is uneven; crystal faces are rare but very large blocky crystals, intergrown with quartz, micas, and other silica-rich minerals, are common. It is brittle, white, gray, pink, buff, or green, is vitreous to dull in luster, and may be opaque to translucent. The hardness is 6 and the specific gravity is 2.54 to 2.57, depending on Na content. It is found in all silica-rich igneous and metamorphic rocks, as in granite batholiths in cores of mountain ranges.

SANIDINE, $KAlSi_3O_8$, is the disordered potassium feldspar. If the ordered feldspar, microcline, is heated to about 900° C, Al^{+3} and Si^{+4} ions diffuse randomly among the tetrahedral positions, forming sanidine. When a rock crystallizes at high temperatures and is rapidly cooled, the disordered form, sanidine, is the observed feldspar. This is commonly the case in lavas, which crystallize rapidly at the surface. Sanidine is monoclinic and is considered to be the high-temperature form. It is notably more glassy in appearance than microcline but crystal habit and all other physical properties are very similar to the other potassium feldspars. Complete characterization is possible only with the aid of X-ray and optical examination. Sanidine is found in silica-rich lavas.

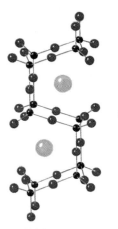

Microcline Crystal Aggregate
Crystal Peak, Colorado

SILICON-OXYGEN FRAMEWORK IN FELDSPARS

Microcline Crystals
Crystal Peak, Colorado

Amazonstone Microcline
Crystal Peak, Colorado

Pink Microcline
Parry Sound, Ontario

Pink Microcline
Keystone, S. Dakota

White Microcline
Keystone, S. Dakota

Amazonstone Microcline
Ontario

**Sanidine, Large Crystals
in Trachyte Lava**
Drachenfels, West Germany

Microcline in Syenite
Wausau, Wisconsin

ORTHOCLASE AND ADULARIA, KAlSi$_3$O$_8$, are two forms of potassium feldspar in which there appears to be partial ordering among the tetrahedral ions but not complete ordering as in microcline. Orthoclase and adularia are the results of rather special crystallization conditions, particularly when water is present at low temperatures (25°–400° C). The name orthoclase is generally reserved for monoclinic crystals with well-formed faces, commonly twinned, in which some portions of the crystals (domains) are ordered and some are not. The domains are not visible to the unaided eye but must be studied by X-ray methods. The name adularia is used for monoclinic crystals having a distinctive rhombic cross-section, with ordered domains large enough to be seen with microscope.

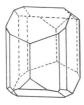 Orthoclase is a fairly common mineral found in shallow intrusive igneous rocks, particularly in pegmatites, as a low temperature precipitate in sedimentary rocks, and as a late stage precipitate in hydrothermal veins. Adularia is restricted generally to vein deposits where it occurs with vein quartz and the sulfides. The physical properties of both are practically identical with those of microcline. In the absence of the distinctive crystal forms common for orthoclase and adularia, it is best to refer to the mineral simply as a potassium feldspar, or K-spar, until optical or X-ray examination has been made.

PERTHITE, KAlSi$_3$O$_8$ and Na AlSi$_3$O$_8$, is an intergrowth of two minerals, microcline and anorthite. At the temperatures at which most feldspars are crystallized, the two minerals are completely soluble, but on cooling the solubility limits are decreased and the two minerals unmix (exsolve) by diffusion through the solid. The Na^{+1} ions will aggregate in bands oriented along certain crystallographic directions with K^{+1} ions migrating from those bands. The result is a single crystal with alternating bands of microcline and albite. If the microcline bands occupy the majority of the crystal volume, it is called a perthite. If albite occupies the greater portion, it is called an antiperthite. Most microclines are perthitic.

ANORTHOCLASE, (K, Na) AlSi$_3$O$_8$, is a solid solution of potassium and sodium feldspar, formed by cooling rapidly so that exsolution could not take place to form a perthite. The solid solution persists indefinitely at normal temperatures. It occurs only in lavas and may be indistinguishable from sanidine without chemical analysis or X-ray study. It is found at Obsidian Cliff in Yellowstone National Park and other similar geologic settings.

CELSIAN, Ba Al$_2$Si$_2$O$_8$, is a feldspar with the Ba^{+2} ion in interstitial positions. Because barium is rare, the mineral is rare, though Ba^{+2}, as well as Sr^{+2} and Pb^{+2}, are commonly found in small amounts in the common feldspars. Potassium feldspars with Ba^{+2} in measurable amounts has been called hyalophane.

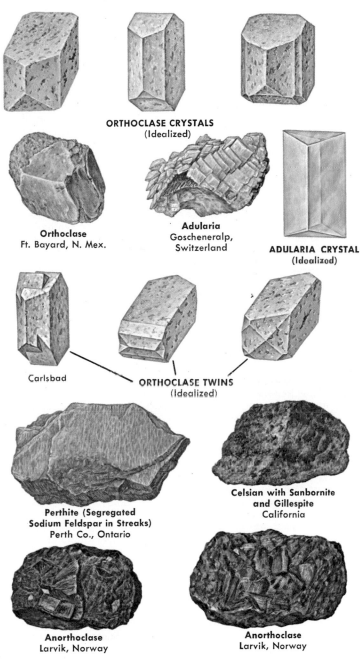

ORTHOCLASE CRYSTALS
(Idealized)

Orthoclase
Ft. Bayard, N. Mex.

Adularia
Goscheneralp,
Switzerland

ADULARIA CRYSTAL
(Idealized)

Carlsbad

ORTHOCLASE TWINS
(Idealized)

**Perthite (Segregated
Sodium Feldspar in Streaks)**
Perth Co., Ontario

**Celsian with Sanbornite
and Gillespite**
California

Anorthoclase
Larvik, Norway

Anorthoclase
Larvik, Norway

THE PLAGIOCLASE SERIES, (Na,Ca) (Al,Si)$_4$O$_8$, is a complete solid solution series. Because the Na^{+1} and Ca^{+2} ions are nearly the same size, a plagioclase can have a composition of any value between albite, NaAlSi$_3$O$_8$, and anorthite, Ca Al$_2$Si$_2$O$_8$. Because of the difference in ionic charges, however, the Al/Si ratio varies with the Ca/Na ratio to maintain electrical neutrality. Mutual substitution of Ca^{+2} and Al^{+3} for Na^{+1} and Si^{+4}, because of the slightly different sizes and different charge distributions, causes slight variations in all properties, most notably specific gravity and cleavage angles. The structure of the plagioclase feldspars is essentially the same as for potassium feldspars.

The series is commonly divided arbitrarily into compositional ranges, as shown on page 219, with a different name applied to each range. The physical properties of all are similar to those of the potassium feldspars; one notable feature of plagioclase feldspars, however, is the common occurrence of polysynthetic twinning. At closely spaced intervals the orientation of the growing crystal changed so that the whole crystal is like a series of layers, chemically bonded together across boundaries, called composition planes. These planes are visible as striations on the surface of the crystal and in many cases cause interference bands to appear in reflected light, causing a blue glow, called chatoyancy. Certain labradorite specimens are particularly prized because of the chatoyancy.

CRYSTALLIZATION of the plagioclase feldspars from a silicate melt is moderately complex and important to the origin of the igneous rock types. Pure anorthite, CaAl$_2$Si$_2$O$_8$, melts at about 1550° C and pure albite, NaAlSi$_3$O$_8$, melts at about 1100° C. A melt containing both Na and Ca, however, crystallizes with precipitation of the plagioclase feldspars in a manner typical of complete solid solution series. The initial crystals formed will be much richer in Ca than the liquid itself. As the temperature is lowered, the crystal reacts with the remaining liquid—Na^{+1} ions diffuse into the crystal—so that the composition of the crystal becomes more Na-rich. If no separation occurs during crystallization, the final crystals will have the same composition as the original melt. If separation occurs, however, e.g. by settling of Ca-rich feldspars before they have time to react with the liquid, a series of different feldspars, from Ca-rich through Na-rich, will be formed. If cooling is very rapid, as in a lava, insufficient time for reaction will cause formation of zoned feldspar crystals, with Ca-rich material in the center and bands of increasingly Na-rich material outward from the center. Separations of plagioclase feldspars of different compositions at different times during the cooling of magma is one mechanism that makes possible the formation of many igneous rock types from a common, homogeneous magma. As shown by Bowen's reaction series (pp. 82-83) similar reactions occur in the ferromagnesian minerals, with olivine forming, then changing to pyroxene, then to amphibole, then to biotite, as cooling occurs.

THE PLAGIOCLASE SERIES

Albite	100-90% $NaAlSi_3O_8$
Oligoclase	90-10 to 70-30
Andesine	70-30 to 50-50
Labradorite	50-50 to 30-70
Bytownite	30-70 to 10-90
Anorthite	90-100% $CaAl_2Si_2O_8$

ALBITE CRYSTALS
(Idealized)

Albite
Bancroft, Ontario

Albite, Twinning Lamellae
(Peristerite)
Cardiff Twp.,
Ontario

Albite
Keystone,
S. Dak.

Oligoclase
Mitchell Co., N.C.

Oligoclase
Showing Lamellae
Kragero, Norway

Oligoclase
(Sunstone)
from Pegmatite
Tvedestrand, Norway

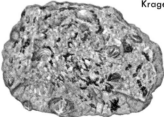

Andesine in Monzonite
Judith Basin, Mont.

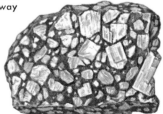

Zoned Plagioclase
in Trachyte
Bannockburn Twp.,
Ontario

PLAGIOCLASE FELDSPARS are found in all igneous rocks except the most silica-deficient, the ultrabasic rocks. They are common as metamorphic minerals, formed by solid reactions in shales at high temperatures and pressures. They weather more rapidly than the potassium feldspars and so are less common in sedimentary rocks but are found in graywackes—sandstones formed from poorly weathered igneous and metamorphic source rocks.

The plagioclase feldspars are difficult to distinguish from the potassium feldspars without detailed optical or X-ray investigation. The most distinctive feature is the striated cleavage surface resulting from twinning. These striations may be absent, however. The only clue as to plagioclase composition that may be obtained without analysis is the association with other minerals.

The properties listed below are for the two end members, albite and anorthite. The intermediate compositions have intermediate properties.

ALBITE, Na $AlSi_3O_8$, is triclinic, with two directions of cleavage at about 86° and one poor cleavage in third direction. The fracture is uneven. Small crystals similar to microcline; tabular crystals, commonly elongate, are found. It is normally intergrown with quartz and micas and crystals are therefore irregular in shape. Polysynthetic twinning is common. Pure $KAlSi_3O_8$ is colorless but minerals may be gray; luster is vitreous, with opalescence observed (moonstone). It is transparent to translucent. The specific gravity is 2.63, increasing with greater Ca content. Albite fuses to a colorless glass; colors flame yellow (Na). Albite weathers to kaolinite clay under basic conditions. It occurs worldwide in igneous and metamorphic rocks of high silica content. It is a constituent of perthite, with microcline. More Ca-rich plagioclases—oligoclase and andesine—may be more highly colored than albite because of impurities. Pure $NaAlSi_3O_8$ is almost nonexistent; several percent Ca is nearly always present.

ANORTHITE, $CaAl_2Si_2O_8$, is triclinic, with two directions of cleavage at about 86° and poor cleavage in the third direction. The fracture is uneven. Crystals are commonly prismatic or tabular; crystals are normally intergrown with amphiboles and pyroxenes, so crystal faces are rare. Polysynthetic twinning is common. The hardness is 6; the specific gravity of pure $CaAl_2Si_2O_8$ is 2.76, decreases with increasing Na content. The pure compound is colorless but minerals may be white, gray, or bluish. The luster is vitreous, occasionally pearly; crystals are transparent to translucent. Anorthite fuses with greater difficulty than albite. It is rapidly decomposed by hydrochloric acid with formation of silica gel. Anorthite weathers rapidly to bauxite and clays. It occurs in basic igneous rocks—gabbros and basaltic lavas—in metamorphic rocks, and much less commonly in sediments. Crystals have been found at Mt. Vesuvius, in Japan, Iceland, Rumania, in the Ural Mountains, and at Franklin, New Jersey.

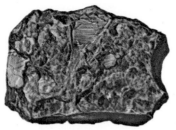

Labradorite
Lake St. John, Quebec

Chatoyant Labradorite
Nain, Labrador

**Labradorite Phenocrysts
in Diabase**
Cape Ann, Mass.

Labradorite
Essex Co., N.Y.

Bytownite
Crystal Bay, Minn.

Bytownite
Crystal Bay, Minn.

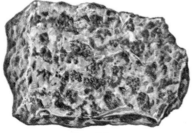

Anorthite
Grass Valley, Calif.

**POLYSYNTHETIC TWINNING
IN ALBITE**
(Idealized)

221

FELDSPATHOID GROUP is often considered to include only nepheline and leucite, but is here used as a general name for the framework silicates containing less silica than the feldspars. The feldspathoids are formed in environments of low silica content and are thus not found where quartz is abundant. They are associated with silica-poor rocks, particularly those rich in soda and potash, such as syenites and trachytes.

NEPHELINE, $Na_3KAl_4Si_4O_{16}$ (sodium potassium aluminum silicate), has a structure essentially like that of tridymite (p. 212), but with Al^{+3} in half the tetrahedral positions and the alkali ions Na^{+1} and K^{+1} in the open space of the framework. The formula approximates the composition of the most common specimens, but some have less K^{+1} or none at all, and others have considerably less than half the Si^{+4} replaced by Al^{+3}. If Na^{+1} is absent, the mineral has a formula simplified to $KAlSiO_4$ and is called kalsilite. Nepheline is a widespread mineral; it is found in silica-poor rocks, associated with the alkali feldspars, and in lavas, particularly basalts. It is commonly associated with other feldspathoid minerals. The greasy luster is diagnostic. Nepheline syenite is an important rock material for use in the manufacture of ceramic whiteware bodies and glazes.

Nepheline crystallizes in hexagonal system as prismatic crystals, disseminated grains, and masses. It is brittle; colorless, white, or gray; greasy or glassy in luster. It has poor cleavage and subconchoidal fracture; normally contains some calcium; is gelatinized by acids; is altered by water to other feldspathoids or to zeolites. Hardness is 5.5-6, specific gravity 2.56-2.67. Occurs notably in Maine, Arkansas, Ontario, Greenland, U.S.S.R., Rumania, Norway, Italy (in lavas of Mt. Vesuvius), and Japan.

EUCRYPTITE, $LiAlSiO_4$ (lithium aluminum silicate), is interesting because, like nepheline, it has a silica structure with Al^{+3} in half the tetrahedral positions and with a $^{+1}$ ion in interstitial positions. Because Li^{+1} is smaller than Na^{+1} or K^{+1}, however, eucryptite assumes the quartz structure (p. 208) rather than the tridymite structure (p. 214). An uncommon mineral, it is apparently formed by alteration of spodumene, $LiAlSi_2O_6$, as at Branchville, Conn. It is commonly formed in lithium-bearing ceramic products with low thermal expansion. Colorless or white, it crystallizes in hexagonal system, with specific gravity of 2.67.

CANCRINITE, $(Na,Ca)_{7-8}Al_6Si_6O_{24}$-$(CO_3,SO_4,Cl)_{1.5-2} \cdot 1-5H_2O$ (hydrous sodium-calcium aluminum silicate), is an igneous mineral formed during later stages of crystallization, probably by hydrothermal alteration of nepheline. Proportions of carbonate, sulfate, chloride, and water vary. It crystallizes in hexagonal system, normally in massive form. It is colorless, white, yellow, reddish, blue, or gray; glassy or greasy in luster. Cleavage is prismatic; hardness 5-6, specific gravity 2.51-2.42. Effervesces in hydrochloric acid. Found with nepheline and sodalite in Maine, Ontario, U.S.S.R., and Norway.

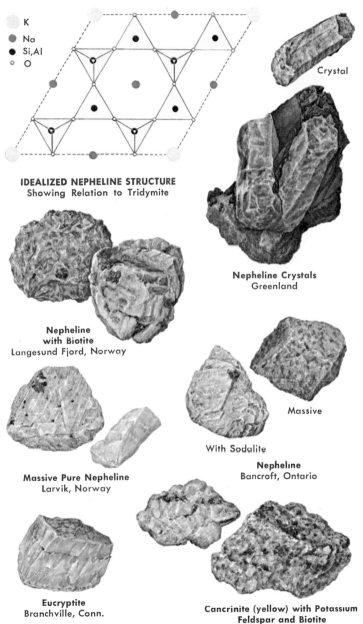

K
Na
Si, Al
O

IDEALIZED NEPHELINE STRUCTURE
Showing Relation to Tridymite

Crystal

Nepheline Crystals
Greenland

**Nepheline
with Biotite**
Langesund Fjord, Norway

Massive

With Sodalite

Nepheline
Bancroft, Ontario

Massive Pure Nepheline
Larvik, Norway

Eucryptite
Branchville, Conn.

**Cancrinite (yellow) with Potassium
Feldspar and Biotite**
Litchfield, Maine

223

SODALITE, $Na_8(Al_6Si_6O_{24})Cl_2$ (sodium aluminum silicate with chlorine), is relatively rare, forming only in silica-poor igneous rocks with high soda content in the presence of chlorine-bearing waters, or in some contact metamorphic limestones. Other silica-poor minerals, notably leucite and nepheline, are associated with it. As with other framework silicates, the composition varies widely, and a chemical analysis is necessary to identify the mineral correctly. Pure sodalite is colorless, but impurities may color it white, blue, green, pink, or yellow.

Sodalite crystallizes in cubic system as dodecahedral crystals, masses, and concentric cavity fillings. It has no well-developed cleavage. Hardness is 5.5-6, and specific gravity is less than that of the feldspars—2.27-2.33, depending on composition. Many specimens show fluorescence in ultraviolet light. Found as crystals in lavas of Mt. Vesuvius and in small amounts in nepheline syenites in many areas.

NOSELITE, $Na_8(Al_6Si_6O_{24})SO_4$ (sodium aluminum silicate with sulfate), has properties very similar to those of sodalite. Less common, it occurs in silica-poor lavas and volcanic debris along the lower Rhine, in the Cape Verde Islands, and in Cornwall, England. Only in northern China has it been found in intrusive igneous rocks.

HAUYNITE, $(Na,Ca)_{4-8}(Al_6Si_6O_{24})$-$(SO_4,S)_{1-2}$ (sodium-calcium aluminum silicate with sulfate-sulfur), is best known as deep-blue, semiprecious gemstone, lazurite (lapis lazuli). Synthetic crystals are generally called ultramarine. Occurs in silica-poor, soda-rich igneous rocks and metamorphosed limestones. Properties similar to sodalite's.

LEUCITE, $KAlSi_2O_6$ (potassium aluminum silicate), has a low silica content and is found only in potassium-rich igneous rocks, either lavas or shallow intrusives. Relatively rare, it occurs notably in Magnet Cove, Ark.; Leucite Hills, Wyo.; Yellowstone Nat. Park; Vancouver Island, B.C.; Italy; and Central Europe.

Leucite crystallizes in tetragonal system as nearly trapezohedral crystals and disseminated grains. It is white or gray, opaque to translucent, vitreous in luster. Commonly contains magnetite or glass inclusions. Cleavage is poor; hardness 5-5.6; specific gravity 2.47-2.50.

ANALCITE, $NaAlSi_2O_6 \cdot H_2O$ (sodium aluminum hydrous silicate), is similar to leucite, but contains water and occurs more commonly with zeolite minerals. It forms in igneous rocks either by crystallization from a melt or by precipitation from water into vesicles (bubbles) in lava. It also occurs in sedimentary rocks, sandstones, phosphate beds, coal deposits, clay deposits formed from volcanic ash, and hot springs and geysers. It crystallizes in cubic system as trapezohedra, cubes, and massive granular deposits. It is generally colorless or white, translucent to transparent, vitreous in luster. Has poor cleavage, subconchoidal fracture. Hardness is 5-5.5, specific gravity 2.24-2.29.

224

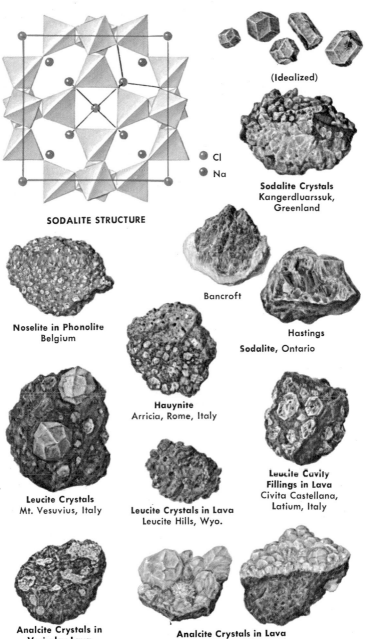

(Idealized)

Sodalite Crystals
Kangerdluarssuk,
Greenland

SODALITE STRUCTURE

Cl
Na

Bancroft

Noselite in Phonolite
Belgium

Hastings

Sodalite, Ontario

Hauynite
Arricia, Rome, Italy

Leucite Crystals
Mt. Vesuvius, Italy

Leucite Crystals in Lava
Leucite Hills, Wyo.

**Leucite Cavity
Fillings in Lava**
Civita Castellana,
Latium, Italy

**Analcite Crystals in
Vesicular Lava**
Tick Canyon, Calif.

Analcite Crystals in Lava
Golden, Colo.

225

SCAPOLITE is a group of minerals whose compositions form a series between meionite, $Na_4Al_3Si_9O_{24}Cl$, and marialite, $Ca_4Al_6Si_6O_{24}CO_3$, neither of which has been found in nature. Scapolite is common in metamorphosed limestones, appearing as distinctive gray grains with glassy luster in white calcite marble, and is less common in other metamorphic rocks. It generally forms at high temperatures and pressures. Also called wernerite, it is common in areas of regional metamorphism, as in New York and New Jersey, throughout Scandinavia, and in many localities in Ontario and Quebec. A yellow variety that is found in Madagascar has been used as a gemstone.

Scapolite crystallizes in tetragonal system, generally as granular inclusions or massive aggregates in marbles, rarely as blocky, pyramidal crystals. Prismatic cleavage is visible, and luster is characteristically vitreous or pearly. Hardness is 5-6, specific gravity 2.50-2.78 (depending on composition). The names mizzonite, dipyre, and passonite have been applied to scapolites.

ZEOLITE GROUP includes 22 well-defined species, of which three common minerals are described here. They are hydrated aluminum silicates and can be dehydrated and rehydrated without destroying their structure. If these minerals are surrounded by water, the sodium in them can be replaced by calcium, or the calcium in them by sodium, if calcium or sodium is present in sufficient concentration in the water.

STILBITE, $(Ca,Na_2,K_2)Al_2Si_7O_{18} \cdot 7H_2O$ (hydrous calcium-sodium-potassium aluminum silicate), occurs as cavity fillings in basaltic lava and as crack fillings in many other rocks. It commonly occurs as distinctive radiating clusters and sheaflike aggregates.

Stilbite crystallizes in monoclinic system and has perfect cleavage in one direction. It is generally white, less commonly yellow, red, or brown. Hardness is 3.5-4, specific gravity 2.1-2.2. Found notably at Paterson, N.J., and in Nova Scotia, Iceland, Scotland, and India.

CHABAZITE, $CaAl_2Si_4O_{12} \cdot 6H_2O$ (hydrous calcium aluminum silicate), is found as fillings in crevices and vesicles in basaltic and other lavas. Properties are similar to those of other zeolites, but it crystallizes in rhombohedral system as white crystals, commonly penetration twins. Composition may vary over wide range, as with most zeolites.

HEULANDITE, $(Ca,Na_2)Al_2Si_7O_{18} \cdot 6H_2O$ (hydrous calcium-sodium aluminum silicate), is similar to stilbite, but occurs as globular and granular crystals in lava cavities. It is also found as a product of devitrification and hydration of volcanic glass. It occurs in extensive bedded deposits in New Zealand and in small quantities in other areas.

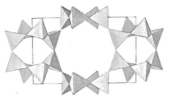

SILICA FRAMEWORK OF SCAPOLITES
Na, Ca, Cl, CO₃, SO₄
occupy positions in holes.

SCAPOLITE CRYSTAL
(Idealized)

Green Scapolite
Otter Lake, Quebec

Massive Scapolite
Homesteadville, N.Y.

**Typical Zeolite Cavity
Filling in Lava**

Stilbite Crystals
Nova Scotia

Chabazite
Wassons Bluff,
Nova Scotia

Chabazite Crystals
Melbourne, Australia

Heulandite
San Juan Co., Colo.

Chabazite Crystals
West Paterson, N.J.

Scotland   New Jersey
Heulandite Crystals

CARBONATES

Carbon dioxide in the air dissolves in water with formation of carbonic acid, consisting of hydrogen ions, H^{+1}, and bicarbonate ions, $(HCO_3)^{-1}$, in solution. If conditions of temperature, pressure, and concentration are right, metal ions will combine with the bicarbonate ions to form carbonates, inorganic salts of carbonic acid. Carbonate structures consist of triangular $(CO_3)^{-2}$ ions held together by metal ions. Of the 70 or so carbonate minerals, calcite and dolomite are the only very abundant species.

CALCITE GROUP has a structure similar to that of halite (p. 120), with the $(CO_3)^{-2}$ ions in the Cl^{-1} positions, the Ca^{+2} ions in the Na^{+1} positions, and the cubic structure compressed along one axis. Each Ca^{+2} ion is surrounded by six oxygen atoms of the CO_3 groups. The relationship between structure and cleavage is shown on the facing page. Like NaCl, calcite has perfect cleavage in three directions but not at right angles. There is extensive solid solution (p. 32) among the members of the calcite group.

CALCITE, $CaCO_3$ (calcium carbonate), is one of the commonest minerals, being the main constituent of limestone. It is formed from carbonic acid (derived from solution of atmospheric CO_2) and Ca^{+2} ions (dissolved from continental rocks). These are extracted from water by many organisms, notably certain algae and invertebrates, and made into calcium carbonate skeletons. The skeletons, plus calcium carbonate directly precipitated from the water, form reefs and associated limestone beds. Most calcite occurs as small crystals cemented into massive limestones. Heat and pressure can cause recrystallization and grain growth, as in the metamorphic rock, marble. Large crystals also grow from solution in cavities, as in hydrothermal veins and caverns. The cleavage surfaces do not coincide with common crystal faces.

Calcite crystallizes in rhombohedral system. Dogtooth spar crystals are common, but other shapes occur. It is soft (hardness 3), and specific gravity is 2.71. In limestone, calcite is white or gray, fine-grained, and massive, but crystals may be colorless or tinted yellow, blue, green, or pink. Even in minute amounts, calcite can be detected by applying hydrochloric acid, which causes effervescence.

SIDERITE, $FeCO_3$ (iron carbonate), has the calcite structure, with Fe^{+2} rather than Ca^{+2} in metal positions. It is found in bedded sedimentary iron ores and as a vein deposit with other metal ores. It is a major iron-ore mineral in clay ironstones of England and many other areas. It is harder (3.5-4.5) and heavier (specific gravity 3.48-3.96) than calcite. It generally occurs in shades of brown, but may be green, gray, or black. It oxidizes and hydrolyzes easily to form limonite.

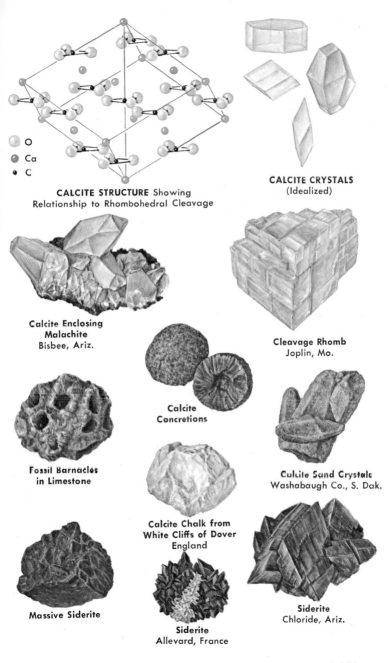

CALCITE STRUCTURE Showing Relationship to Rhombohedral Cleavage

CALCITE CRYSTALS
(Idealized)

O — O
Ca
C

Calcite Enclosing Malachite
Bisbee, Ariz.

Cleavage Rhomb
Joplin, Mo.

Calcite Concretions

Fossil Barnacles in Limestone

Calcite Sand Crystals
Washabaugh Co., S. Dak.

Calcite Chalk from White Cliffs of Dover
England

Massive Siderite

Siderite
Allevard, France

Siderite
Chloride, Ariz.

DOLOMITE, CaMg(CO$_3$)$_2$ (calcium magnesium carbonate), is structurally similar to calcite, but with Mg^{+2} and Ca^{+2} in alternating metal positions. For this reason, it is sometimes put in a separate group. Dolomite may precipitate directly from solutions or may be formed by action of Mg-rich solutions on calcite. Dolomite rocks are present as sedimentary layers in many areas and as products of regional metamorphism of carbonate rocks elsewhere. In appearance it is virtually indistinguishable from calcite, but is less soluble in hydrochloric acid, effervescing only slightly. Crystals are formed in hydrothermal veins, associated with other Mg minerals. Dolomite is used as an ornamental stone and as a raw material for manufacture of cement and of magnesium oxide for refractories. Because of the size difference between the Ca^{+2} and Mg^{+2} ions, complete solid solution with calcite does not occur.

Dolomite crystallizes in rhombohedral system, forming crystals like those of calcite, but also with curved faces and complex growths. Cleavage is perfect in 3 directions. Luster is vitreous or pearly, and color may be white, yellow, brown, red, green, gray, or black. Massive dolomite is white to gray. Properties very similar to calcite's. Common in coarse marble deposits in areas of regional metamorphism.

MAGNESITE, MgCO$_3$ (magnesium carbonate), is formed by action of Mg-rich solutions on rocks, by alteration of calcite or dolomite, or by carbonation of silicate rocks. Properties are similar to those of calcite, but magnesite is harder (3.75-4.25) and heavier (specific gravity about 3.48) and effervesces in hot (but not cold) hydrochloric acid. Major deposits are found in montainous areas of western U.S. and Europe and on the Island of Euboea, Greece. It is a source of MgO for industry.

ANKERITE, Ca(Mg,Fe,Mn)(CO$_3$)$_2$ (calcium magnesium-iron-manganese carbonate), has been called a calcium iron carbonate, but considerable magnesium and manganese are always present in the natural material. Properties and occurrence are essentially the same as those of dolomite; in fact, ankerite represents nothing more than a chemical variety of dolomite. Found with other iron-bearing minerals in Austria and other areas in the Alps, in Mexico at Guanajuato, and in New York, Nova Scotia, and elsewhere.

RHODOCHROSITE, MnCO$_3$ (manganese carbonate), is a favorite of mineral collectors because of its rose-red color, though it is too soft to be valuable as a gemstone. Properties and occurrence are very similar to those of siderite (p. 228), and extensive solid solution occurs with that mineral. Color may be yellow, brown, gray, or dark red, depending on composition. Rhodochrosite is relatively rare; pure specimens are found as gangue material in veins with other Mn minerals.

SMITHSONITE, ZnCO$_3$ (zinc carbonate), is formed by action of carbonated water on zinc sulfide or by alteration of carbonate rocks by zinc solutions. It is commonly associated with hemimorphite (calamine) and with lead-zinc deposits. Porous variety is called dry-bone ore. It occurs as crusts, stalactites, or botryoidal masses in veins or beds. It is much harder (5.5) than calcite and much heavier (specific gravity 4.30-4.45). Generally white, it may be gray, green, brown, or blue. Luster is glassy or pearly.

Dolomite Crystals
Ossining, N.Y.

Dolomite
(Magnesia Marble)
Ossining, N.Y.

Dolomite, Twin Crystals
Grisons, Switzerland

Magnesite (Massive)
Chewelah, Wash.

Dolomite Crystals
with Pyrite
Joplin, Mo.

Rhodochrosite
Colorado

Magnesite
Greece

Magnesite Crystal
(var. Breunerite)
Pfitsch, Arizona

Rhodochrosite
Argentina

Ankerite Crystals
Marlboro, Vt.

Smithsonite
Leadville, Colorado

Ankerite and Quartz
Phoenixville, Pa.

Smithsonite
Chihuahua, Mexico

Smithsonite Crystals
Tsumeb, South-West Africa

231

ARAGONITE GROUP of minerals differs in structure from the calcite group (p. 228) in that the metal ions are shielded by 9 rather than 6 oxygen ions. The additional shielding is required by the larger ions of barium, strontium, and lead, which is why the carbonates of these ions are found only with the aragonite structure. But calcium carbonate, with its smaller calcium ion, may crystallize either as calcite (under ordinary conditions) or as aragonite. Arrangement of Ca^{+2} and $(CO_3)^{-2}$ ions is nearly hexagonal, hence the crystals are commonly pseudohexagonal.

ARAGONITE, $CaCO_3$ (calcium carbonate), can be formed only under certain poorly understood conditions, as in cavern deposits. At ordinary temperatures and pressures it is unstable, changing to calcite in hours to years. Many organisms that construct calcium carbonate shells can form aragonite; in fact, some have shells that are aragonite in one part and calcite in another. When an organism discards such a shell, the aragonite normally changes to calcite, but its characteristic needlelike structure is retained in the crystal shape. Aragonite can be converted to calcite at ordinary temperatures by grinding it to a fine powder.

Aragonite crystallizes in orthorhombic system, but forms cyclic twins with a hexagonal section. It has 2 directions of cleavage and glassy luster. It occurs in groups of radiating crystals or in massive deposits. It is common as deposits in caverns, hot springs, and cavities of lava flows, and as precipitates associated with gypsum beds. Color is normally white, but some varieties are gray, green, yellow, or violet. Hardness is 3.5-4, specic gravity 2.94-2.95.

WITHERITE, $BaCO_3$ (barium carbonate), is associated with barite, the other important but more common barium mineral, in low-temperature hydrothermal veins in which galena is the major ore. Witherite can be altered to barite by action of sulfate-bearing water, and barite can be altered to witherite by carbonated water. Witherite crystallizes in orthorhombic system as massive aggregates, columnar aggregates, and nearly hexagonal twinned crystals. It is white, rarely gray or yellow; glassy in luster, and translucent to transparent. Cleavage is in 2 directions. Hardness is 3.5, specific gravity 4.3. Witherite is found in Kentucky, Montana, Illinois, California, Arizona, Ontario, Austria, France, West Germany, Czechoslovakia, U.S.S.R., England, and Japan.

STRONTIANITE, $SrCO_3$ (strontium carbonate), is the strontium analog of witherite. It is associated with celesite (strontium sulfate) in veins or without celestite in veins in limestones. It is similar to witherite in properties, but is most commonly pale green (less commonly white, gray, yellow, or brown) and has specific gravity of 3.72.

CERUSSITE, $PbCO_3$ (lead carbonate), called white-lead ore, is nearly identical to witherite in properties, but has the highest specific gravity—6.5 —of any carbonate. It occurs as an alteration product of anglesite (lead sulfate) and as a precipitate from lead solutions. It alters to hydrocerussite, $Pb_3(CO_3)_2(OH)_2$, basic lead carbonate, and is confused with it. Cerussite effervesces in acid.

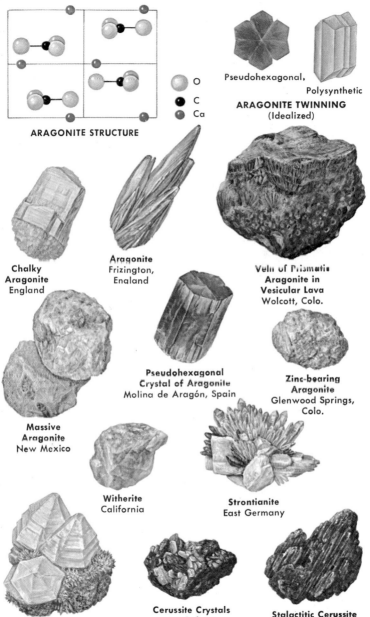

ARAGONITE STRUCTURE

O
C
Ca

Pseudohexagonal,

Polysynthetic

ARAGONITE TWINNING
(Idealized)

Chalky Aragonite
England

Aragonite
Frizington, England

Vein of Prismatic Aragonite in Vesicular Lava
Wolcott, Colo.

Pseudohexagonal Crystal of Aragonite
Molina de Aragón, Spain

Zinc-bearing Aragonite
Glenwood Springs, Colo.

Massive Aragonite
New Mexico

Witherite
California

Strontianite
East Germany

Witherite Crystals
Fallowfield, Northumberland, England

Cerussite Crystals on Galena
Coeur d'Alene, Idaho

Stalactitic Cerussite
Joplin, Mo.

233

MISCELLANEOUS CARBONATES

AZURITE, $Cu_3(CO_3)_2(OH)_2$ (basic copper carbonate), is so named because of its blue color, which also makes it a popular collector's item. The color is distinctive, though the shade of blue varies among specimens. Azurite occurs only in the oxidized portions of copper-ore veins. It is formed by action of carbon dioxide and water on copper sulfides or by action of copper solutions on calcite. It was once an important source of copper, but the near-surface oxidized zones where it occurs have been largely mined, and the deeper copper sulfide ores are now used.

Azurite crystallizes in monoclinic system, occurring as tabular or equidimensional transparent crystals with very pure blue color and brilliant luster. Good crystals are relatively rare; azurite is more common as an earthy material closely associated with malachite. It has complex cleavage, specific gravity of 3.77-3.89, and hardness of 3.5-4. It is too soft and cleavable to be a good gemstone.

MALACHITE, $Cu_2CO_3(OH)_2$ (basic copper carbonate), is more abundant than azurite, with which it is found. Like azurite, it forms as an alteration product of copper sulfides or calcite. Malachite has a distinctive green color. (The green weathering product that forms on copper roofs of large buildings is a type of malachite.) Malachite is found in copper mining areas; in some localities, as in the Urals of the U.S.S.R., masses of malachite weighing hundreds of tons have been mined.

Malachite crystallizes in monoclinic system as massive deposits or crusts; as botryoidal, fibrous, or stalactitic masses; or as slender needlelike crystal aggregates. Also, earthy aggregates of azurite and malachite are very common. Except for the green color, malachite's properties are very similar to those of azurite.

NATRON, $Na_2CO_3 \cdot 10H_2O$ (hydrous sodium carbonate), is a water soluble mineral found only as a deposit left by evaporation of water in shallow, undrained lakes in arid regions, as in California, Nevada, and the Middle East. It is always found with other soda compounds, notably trona. At 34.5° C. it loses much of its hydrated water and dissolves in it. Natron crystallizes in monoclinic system and is found as granular crusts or coatings. It is very soft (hardness 1-1.5) and light (specific gravity 1.478).

TRONA, $Na_3H(CO)_2 \cdot 2H_2O$ (hydrous sodium acid carbonate), is another monoclinic water-soluble carbonate. It is formed by evaporation of water from the "soda lakes" of California, Nevada, Egypt, and Venezuela. It is called an "acid" carbonate because of the H^{+1} ion.

HYDROZINCITE, $Zn_5(CO_3)_2(OH)_6$ (basic zinc carbonate), is a secondary mineral formed by action of carbonated water on zinc ores. It is monoclinic in structure and is found as massive, earthy crusts.

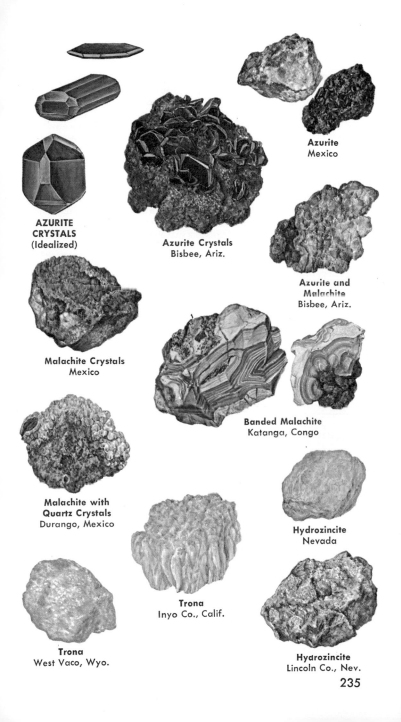

AZURITE CRYSTALS
(Idealized)

Azurite Crystals
Bisbee, Ariz.

Azurite
Mexico

Azurite and Malachite
Bisbee, Ariz.

Malachite Crystals
Mexico

Banded Malachite
Katanga, Congo

Malachite with Quartz Crystals
Durango, Mexico

Trona
Inyo Co., Calif.

Hydrozincite
Nevada

Trona
West Vaco, Wyo.

Hydrozincite
Lincoln Co., Nev.

235

NITRATES

The nitrate ion, NO_3^{-1}, is structurally the same as the carbonate ion, CO_3^{-2}, with one less charge. The complex triangular ion thus joins with the monovalent alkaline earth ions to form crystals with the calcite structure, as in soda-niter, and the aragonite structure, as in niter. Because all the nitrates are very soluble in water, they are not widespread geologically, but are important in the manufacture of two widely used but antithetical products: fertilizer and explosives. There are 10 species of naturally occurring nitrates, only two of which are found in significant amounts and are included here. Others are ammonia niter, NH_4-NO_3; nitrobarite, $Ba(NO_3)_2$; nitrocalcite $Ca(NO_3) \cdot 4H_2O$; nitromagnesite, $Mg(NO_3)_2 \cdot 6H_2O$; gerhardtite and buttgenbachite, complex copper nitrates; and darapskite, a hydrated sodium-nitrate-sulfate. These are not important mineralogically.

NITER, KNO_3 (potassium nitrate), is found in small amounts as a surface coating in arid areas and with soda-niter and other soluble salts as nitrate deposits in Chile and other desert regions. It is commonly associated with organic matter; bacterial action on plant and animal remains generates nitrate, and potassium is extracted from mica-type clays. Niter is found also in limestone caverns, associated with bat guano, notably in Kentucky, Tennessee, and the Mississippi Valley. These deposits were an important source of nitrates used for explosives by the Confederacy when the Union blockade cut off supplies from Chile. Niter is also known as saltpeter.

Niter is orthorhombic in structure, forming needlelike crystals, crusty coatings, and massive or earthy aggregates. It is gray or white, but may be mixed with colored impurities. It is very soluble in water and has a cool salty taste. Hardness is 2, specific gravity 2.1. At 129° C. the structure changes to the rhombohedral calcite type.

SODA NITER, $NaNO_3$ (sodium nitrate), is found with niter, gypsum, halite, iodates, and other minerals in a 450-mile belt along the coast of northern Chile, an area of negligible rainfall. This belt is a few inches to a few feet thick and is covered by a little rock and sand. Nearly all nitrates came from Chile until World War I, when processes were developed for the manufacture of nitrates from atmospheric nitrogen.

Soda niter is rhombohedral in structure, commonly occurring as massive aggregates or crusts on surfaces. It is colorless or white in aggregates and is commonly mixed with many other minerals. It has a hardness of 1.5-2 and a specific gravity of 2.25 if pure. It is very soluble in water and has a cool taste. Small amounts are found in Nevada, California, Bolivia, Peru, Northern Africa, India, and U.S.S.R.

Soda Niter (Wet)
Tarapaca, Chile

Soda Niter
La Noria, Chile

Soda Niter
Tarapaca, Chile

USES OF NITRATES

FERTILIZERS are essential for production of the enormous quantities of food needed to support the increasing human population. Most of the elements necessary for nutrition are available to plants in readily usable form in air, water, and soil. Nitrogen, potassium, and phosphorus, however, may be rapidly depleted from the soil and must be replaced. These are the constituents of commercial fertilizers. Though nitrogen constitutes 80 percent of the atmosphere, only a few kinds of plants can extract it and convert it into usable organic form. Water-soluble nitrate minerals, notably niter and soda-niter (also called saltpeter and Chile saltpeter), were important commodities until recently. Now, by the Haber process, nitrogen can be extracted from the atmosphere for industrial uses as well as for agricultural applications.

EXPLOSIVES have been made from nitrates since the development of gunpowder from niter, carbon, and sulfur in medieval times. Nitrogen can be combined chemically with other substances to form a wide variety of compounds that are unstable. Heat, electrical current, or even a sudden jolt causes them to dissociate, or burn, changing rapidly into gas with a volume many times that of the solid or liquid. Nitroglycerin, $C_3H_5(NO_3)_3$, made by treating glycerin with a mixture of nitric and sulfuric acid, led to the development of dynamite, TNT, and similar explosives that changed the face of the Earth through excavation and war. The essential ingredient, nitric acid, was manufactured exclusively from nitrate minerals and nitrates from animal wastes prior to development of methods for extraction of nitrogen from the atmosphere.

SULFATES

Sulfates are compounds of oxygen and sulfur with one or more metals. The oxygen (O) and sulfur (S) form the sulfate ion, SO_4^{-2}, in which an S atom is surrounded by 4 O atoms located at the corners of a tetrahedron. The 2 excess negative charges are distributed evenly over the O atoms. The sulfates are complex minerals, because the number of ways in which cations can be accommodated among the PO_4^{-2} ions in a crystal structure is large. Over 150 have been named. Many are strongly hydrated species, and many are rare. A few important ones are described. Some of lesser importance are included to illustrate the diversity of the sulfates and some similarities.

GYPSUM, $CaSO_4 \cdot 2H_2O$ (hydrous calcium sulfate), is the commonest sulfate. Huge beds of it occur in sedimentary rock with limestones, shales, sandstones, and clays. Rock salt and sulfur deposits may be associated with it. It occurs as common gypsum and as three varieties having distinctive habits: alabaster, massive; selenite, transparent and foliated; and satinspar, fibrous with silky or pearly luster. Gypsum is by far the most important sulfate because of its use for manufacture of plaster of Paris. In this process the gypsum is heated and loses 75 percent of its water. The resulting hemihydrate (plaster of Paris) readily takes up water and converts back to gypsum. In doing so, its particles recrystallize and become cemented firmly together.

Gypsum crystallizes in monoclinic system, occurring as prismatic or bladed crystals and as fibrous vein-fillings, radiating aggregates, clusters on cave walls, and massive, rock-forming beds. It is generally white or gray, but may be colorless in large crystals, pink in alabaster, or brown and yellow in massive layers. Hardness is 2, specific gravity 2.30. Found in commercial quantities in New York, Michigan, Iowa, Kansas, New Mexico, Colorado, Utah, Canada, and various European countries.

ANHYDRITE, $CaSO_4$ (calcium sulfate), is less abundant than gypsum because it readily takes up water and converts to gypsum. It is deposited by sea water and is therefore found in sedimentary rocks associated with salt beds. Smaller quantities occur in some veins, in caliches (nitrate-bearing beds), and in fumarole deposits. It crystallizes in orthorhombic system, forming usually massive deposits. Crystals are rare. It is colorless if pure, colored if impure. Hardness is 3.5, specific gravity 3. Pearly luster is common.

EPSOMITE, $MgSO_4 \cdot 7H_2O$ (hydrous magnesium sulfate), is more familiar as epsom salts. It is one of many highly hydrated, water-soluble sulfates. It is found in mineral waters and deposits therefrom, in salt lake areas, and as an efflorescence on the walls of caves and mines. It crystallizes in orthorhombic system, occurring as crystals, granular crusts or fibrous aggregates, and as stalactites and botryoidal masses. It commonly contains appreciable quantities of nickel, manganese, zinc, and cobalt.

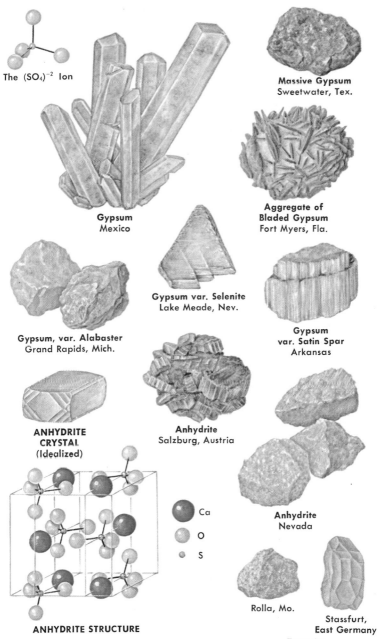

The $(SO_4)^{-2}$ Ion

Massive Gypsum
Sweetwater, Tex.

Gypsum
Mexico

Aggregate of Bladed Gypsum
Fort Myers, Fla.

Gypsum var. Selenite
Lake Meade, Nev.

Gypsum, var. Alabaster
Grand Rapids, Mich.

Gypsum var. Satin Spar
Arkansas

ANHYDRITE CRYSTAL
(Idealized)

Anhydrite
Salzburg, Austria

Ca

O

S

Anhydrite
Nevada

ANHYDRITE STRUCTURE

Rolla, Mo.

Stassfurt,
East Germany

Epsomite

BARITE GROUP includes barite, celestite, and anglesite. These provide an excellent example of how the sizes of ions affect crystal structure, as shown on opposite page. Large Ba^{+2}, Sr^{+2}, and Pb^{+2} ions are packed among the sulfate ions so that each is surrounded by ten oxygen atoms. The barium, strontium, and lead ions are similar in size and, chemically combined with the sulfate ion, form crystals of identical structure. Complete solid solution (p. 32) among the three minerals is therefore structurally possible and has been observed.

BARITE, $BaSO_4$ (barium sulfate), occurs as a common gangue material, particularly in lead-zinc veins in limestone, as in the Joplin, Mo., area. It is also widespread as veins, cavity fillings, and intergranular precipitates in limestones and other sedimentary rocks. It has been found in hot-springs deposits and with some hematite ores. The mineral commonly has significant solid solution with celestite, and properties vary accordingly; solid solution with anglesite is less common. Barite is used as a filler in paints and paper; finely ground, it is mixed with drilling mud to control the mud's specific gravity during oil-well drilling.

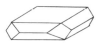 Barite crystallizes in orthorhombic system. Tabular crystals are common, prismatic less so. Barite also occurs as massive aggregates of crystals, globular concretions, and fibrous varieties. Barite roses, radiating clusters of crystals, commonly stained with iron oxides, are collectors favorites. It is generally colorless or white, may be yellow or brown, or may be pigmented by inclusions. Hardness is 3-3.5, specific gravity 4.50.

CELESTITE, $SrSO_4$ (strontium sulfate), is so named because of the distinctive pale blue color of many well-crystallized specimens. It occurs chiefly in sedimentary rocks, with gypsum, anhydrite, and rock salt. Less commonly it is found as gangue material in veins and volcanic rocks. It crystallizes in orthorhombic system as well-developed tabular or elongate crystals with three-dimensional cleavage. It also occurs as fibrous veinlets, radiating aggregates, massive, earthy fine-grained deposits. It is normally white to pale blue, but may be tinted red, brown, or green. Luster is glassy. Earthy varieties may be mixed with clay. Hardness is 3-3.5; specific gravity is 3.97, varying with barium and calcium content. Celestite occurs throughout the world in relatively small quantities.

ANGLESITE, $PbSO_4$ (lead sulfate), is the lead analog of barite and celestite, having the same structure. Its geologic occurrence, however, is quite different. It is almost exclusively an alteration product of galena, PbS, and is found in hydrothermal veins. Cerussite, the lead carbonate, generally forms simultaneously and is mixed with the anglesite. Anglesite crystallizes in orthorhombic system, occurring as well-formed tabular or prismatic crystals. It also is massive, granular, nodular, or stalactitic. Specimens of galena, altered on the surface to anglesite, are common. It is colorless, white, or gray; less commonly yellow, green, or blue. It may be transparent to opaque, depending on particle size and inclusions of cerussite. Hardness is 2.5-3, specific gravity 6.36.

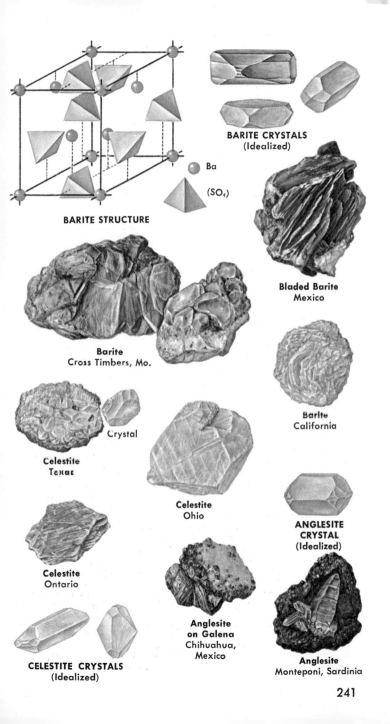

BARITE CRYSTALS
(Idealized)

Ba

(SO₄)

BARITE STRUCTURE

Bladed Barite
Mexico

Barite
Cross Timbers, Mo.

Barite
California

Crystal

Celestite
Texas

Celestite
Ohio

ANGLESITE CRYSTAL
(Idealized)

Celestite
Ontario

Anglesite on Galena
Chihuahua, Mexico

Anglesite
Monteponi, Sardinia

CELESTITE CRYSTALS
(Idealized)

241

BROCHANTITE, $Cu_4(SO_4)(OH)_6$ (basic copper sulfate), is a secondary mineral formed by oxidation of copper sulfide ores, most commonly in arid regions. Though sulfides are normally stable at temperatures and pressures near the earth's surface, the presence of water, oxygen, and acids creates a chemical environment in which they can no longer exist. The oxidation of sulfide ores is thus closely analogous to chemical weathering of igneous or metamorphic rocks. The green color is diagnostic and is typical of the Cu^{+2} ion in the crystal structure.

 Brochantite crystallizes in monoclinic system, forming short prismatic crystals, aggregates of needlelike crystals, granular masses, and crusts. It has perfect one-directional cleavage and fractures unevenly. It is various shades of green, glassy in luster, translucent to transparent, and soluble in acids. Hardness is 3.5-4 (too soft to be used as a gemstone), specific gravity 4.09. Found in many copper mining areas of U.S. (Ariz., Colo., Utah, N. Mex., Idaho, Calif.), Europe, Australia, Africa, and South America.

ANTLERITE, $Cu_3(SO_4)(OH)_4$ (basic copper sulfate), is found with brochantite and is formed in the same way, but has more sulfate relative to copper. Antlerite forms at higher sulfate concentrations, brochantite at lower ones. Antlerite crystallizes in orthorhombic system as tabular crystals or as fibrous aggregates, veinlets, or granular masses. Properties are very similar to brochantite's. Identification is made by study of the crystal forms. Antlerite is found in many copper-mining areas.

GLAUBERITE, $Na_2Ca(SO_4)_2$ (sodium calcium sulfate), is an interesting double salt. It is stable in dry air, but in water it separates into Na_2SO_4 (Glauber salt) and $CaSO_4 \cdot 2H_2O$ (gypsum). Thus it is found only in very arid regions, as in nitrate deposits of Chile, dry lakes of California, and ancient salt deposits of U.S., Europe, U.S.S.R., and India. Glauberite crystallizes in monoclinic system, forming tabular or prismatic crystals embedded in salt deposits or in cavities in basic lavas. It is normally gray or yellow, but is colorless and transparent in unaltered crystals. Hardness is 2.5-3, specific gravity 2.81.

ALUM GROUP is a large and diverse group of chemicals having the general formula $M^{+1}M^{+3}(SO_4)_2 \cdot 12H_2O$, in which M^{+1} is a large ion such as K^{+1}, NH_4^{+1}, Rb^{+1}, or Cs^{+1}, and M^{+3} is a small ion such as Al^{+3} or Fe^{+3}. Naturally occurring alums are combinations of these ions with the sulfate ion. In addition, a number of alums have been synthesized with the selenate ion, $(SeO_4)^{-2}$, in place of the sulfate ion. The alums are isostructural, and extensive solid solution among them is possible. Three alums have been described as minerals: potassium alum, $KAl(SO_4)_2 \cdot 12H_2O$, the commercial alum used in the familiar styptic pencil; ammonia alum, $NH_4Al(SO_4)_2 \cdot 12H_2O$; and the so-called soda alum, $NaAl(SO_4)_2 \cdot 12H_2O$, which is structurally different from the other species. The alums form cubic crystals, occurring as efflorescences in clay-bearing rocks or in coal deposits, as deposits from sulfurous vapors' in volcanic and hot-spring regions, and in some caves. All of the alums are water soluble, have an astringent taste, and lose water readily when heated. Commercial alums are manufactured from alunite, $KAl_3(SO_4)_2(OH)_6$. They are white if pure.

BROCHANTITE CRYSTAL
(Idealized)

Brochantite
Tsumeb, South-West Africa

ANTLERITE CRYSTALS
(Idealized)

Brochantite
Chuquicamata, Chile

GLAUBERITE
(Idealized)

Kalinite (Potassium Alum)
Tonapah, Nevada

Glauberite
Ciempozuelos, Spain

Tschermigite (Ammonia Alum)
Wamsutter, Wyoming

PHOSPHATES, ARSENATES, AND VANADATES

The basic building blocks of these minerals—the phosphate ion, $(PO_4)^{-3}$, the arsenate ion, $(AsO_4)^{-3}$, and the vanadate ion, $(VO_4)^{-3}$ —are tetrahedral, like the sulfate ion (p. 238). The three negative charges on each ion, distributed evenly over the four oxygen atoms, are neutralized by metal ions, forming relatively dense crystals. There are over 240 recognized mineral species, most of them rare. Only a few common minerals are described here. The phosphates are abundant minerals and are important as fertilizers. The arsenates and vanadates are considerably less common, but are important sources of rare elements, notably uranium.

TURQUOISE, $CuAl_6(PO_4)_4(OH)_8 \cdot 4H_2O$ (hydrous basic copper aluminum phosphate), is a popular gemstone because of its blue color and delicate veining. It is formed by alteration of aluminum-bearing surface rocks, with copper derived from weathering of copper sulfides, and phosphates probably dissolved from apatite. It is almost always fine-grained; an iron-rich variety, chalcosiderite, occurs as crusts of fine crystals. Turquoise is associated with limonite, kaolinite, and chalcedony in narrow seams and patches in lavas and some pegmatites. Some natural turquoise specimens with poor color are tinted artificially. Attempts to synthesize turquoise have not been successful. Mixtures of copper and aluminum phosphates, similar in color, are sold as turquoise.

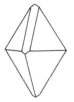 Turquoise crystallizes in triclinic system, usually occurring as dense masses, crusts, and veinlets. These are pale blue, blue-green, green. Rare, minute crystals are transparent bright blue. Other minerals are common as veinlets in best turquoise. Hardness is 5-6, specific gravity 2.6-2.8. Good specimens occur in Arizona, California, Colorado, New Mexico, Virginia, Iran, Siberia, France and West Germany.

LAZULITE, $(Mg,Fe)Al_2(PO_4)_2(OH)_2$ (basic magnesium-iron aluminum phosphate), is not to be confused with lazurite, or lapis lazuli. In contrast to turquoise, lazulite is formed at high temperature in metamorphic rocks, in quartz veins in such rocks, and in granite pegmatites. Some transparent crystals make attractive gemstones. The hardness, 5.5-6, and resistance to cleavage are also good gem qualities. The property that makes lazulite a popular gemstone, however, is the variety of unusual blue colors and the glassy luster.

Lazulite crystallizes in monoclinic system, forming tabular and pyramidal crystals and granular masses. It is azure blue, light blue, dark blue, or blue-green. Streak is white. Luster is glassy. Specific gravity is 3.08-3.38, depending on iron content. Iron-rich varieties are called scorzalite. Good specimens come from Georgia, North Carolina, Vermont, California, New Mexico, Switzerland, Austria, Sweden, Madagascar, Brazil. The color varies with variations in the Mg/Fe ratio.

THE PO₄ ION

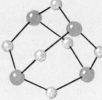

THE P₄O₆ ION

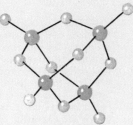

THE P₄O₁₀ ION

● P   ○ O

Turquoise with Quartz
Arizona

Turquoise
Cut En Cabachon

Turquoise
near Cerrillos,
N. Mexcio

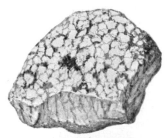

Turquoise (Spider Web)
New Mexico

Lazulite Crystal
Graces Mt.,
Lincoln Co., N.C.

Lazulite Crystals
Salzburg, Austria

Lazulite Crystals
Graves Mt., Georgia

245

APATITE, $Ca_5(PO_4)_3(OH,F,Cl)$ (calcium phosphate with hydroxyl, fluorine, and chlorine), is the commonest phosphate; it occurs as an accessory mineral in igneous rocks (most notably granite pegmatites) and in metamorphic and sedimentary rocks. If rich in OH, it is called hydroxyapatite; in F, fluorapatite; or in Cl, chlorapatite. It is the major constituent of teeth and bones in mammals. Since fluorapatite strongly resists acid attack, fluoride incorporated into teeth during growth helps prevent tooth decay. Because they readily substitute for calcium, radioactive strontium and toxic lead from the environment are absorbed by human bone. Phosphate deposits, derived largely from fossil bone and shells, occur in southeastern U.S. and elsewhere; they are the major source of phosphate fertilizers.

Apatite crystallizes in hexagonal system. It occurs as transparent, glassy, prismatic or tabular crystals or as dull, fibrous or granular aggregates. White if pure, it is more often green, brown, yellow, or blue. Streak is white. Cleavage is poor, fracture conchoidal. Hardness is 5, specific gravity 3.1-3.4. Small amounts are found worldwide.

MONAZITE, (Ce,La,Th)PO₄ (cerium-lanthanum-thorium phosphate), may contain uranium as well as Ce, La, and Th, and thus constitutes the major source of these increasingly important raw materials. It is found as an accessory mineral in granitic igneous rocks; because of its high specific gravity, 5.0-5.3, it is commonly segregated by stream action and concentrated in river and beach sands as placer deposits. Commercial deposits exist in Brazil, India, Malaysia, New Zealand, and southeastern U.S. Monazites containing uranium and thorium are used to determine ages of igneous rocks by isotope analysis. Monazite crystallizes in monoclinic system. It occurs as small, equidimensional or elongated crystals with well-developed faces, distributed sparsely through granites, and as large crystals in some pegmatites. Monazite sands are fine-grained, with rounded crystals. Crystals generally translucent. Color red or brown, luster resinous, cleavage one-directional, fracture conchoidal or uneven, and hardness 5-5.5. Also commonly contains yttrium, neodymium, and other lanthanides.

ERYTHRITE, $(Co,Ni)_3(AsO_4)_2 \cdot 8H_2O$ (hydrous cobalt-nickel arsenate), is similar to vivianite in structure, method of formation, and occurrence; it illustrates chemical similarity of phosphates and arsenates. Contains more Co than Ni; **annabergite** is same mineral but with more Ni than Co. Color is reddish.

VIVIANITE, $Fe_3(PO_4)_2 \cdot 8H_2O$ (hydrous iron phosphate), is a secondary phosphate formed by alteration of ore deposits near the surface or of primary phosphates in pegmatites. Occurs as small prismatic crystals (monoclinic system) or as globular masses and crusts. May be colorless, blue, or green.

VARISCITE, $(Al,Fe)PO_4 \cdot 2H_2O$ (hydrous aluminum-iron phosphate) is formed by direct deposition from phosphate-bearing water that has reacted with Al-rich rocks in near-surface environments. Occurs as fine-grained masses in nodules, cavity fillings, and crusts. Orthorhombic, glassy, white to greenish. Strengite is iron-rich variety. Complete solid solution occurs.

Apatite Crystals
Durango, Mexico

APATITE CRYSTALS
(Idealized)

Massive Apatite
Wilberforce, Ontario

Apatite Crystals
Renfrew, Ontario

Apatite with Biotite
North Carolina

Arendal

Iveland

Monazite Crystals
Norway

Vivianite
Victoria, Australia

Erythrite
Cobalt, Ontario

Monazite Sand
Travancore, India

Erythrite
Schneeberg,
East Germany

Monazite
San Miguel Co., N. Mex.

Variscite
Lewiston, Utah

247

CARNOTITE, $K_2(UO_2)_2(VO_4)_2 \cdot nH_2O$ (potassium uranium vanadate), is the chief ore of uranium, important in atomic-energy applications. Chemistry is very similar to that of phosphates and arsenates. Carnotite is the result of alteration of vanadium- and uranium-bearing minerals by water near the surface. It occurs as a distinctive yellow or greenish yellow crust or cavity filling and is found incorporated into petrified wood. Major deposits exist in Colorado Plateau region of U.S. (Colorado, Utah, New Mexico, and Arizona), most commonly as intergranular material in sandstones and in petrified wood; also at Radium Hill, South Australia; in Katanga Province, the Congo; and in Canada. Uraninite, UO_2, recovered from pitchblende, in which it occurs with radium, is also a major source of uranium.

Carnotite crystallizes probably in orthorhombic system (structure has not been well-defined). It is very finely crystalline and always occurs as soft aggregates. Water content varies with humidity. Crystals appear to have one-directional platy cleavage. Luster is generally dull or earthy, but larger crystals appear pearly. Specific gravity is near 5. Occurs with tyuyamunite and several other vanadate minerals.

TYUYAMUNITE, $Ca(UO_2)_2(VO_4)_2 \cdot nH_4O$ (calcium uranium vanadate), appears to be structurally different from carnotite, but is similar in its chemistry and its distinctive platy cleavage and shades of yellow color. Like carnotite, it is formed by action of water on uranium- and vanadium-bearing minerals, but contains calcium instead of potassium. Whether carnotite or tyuyamunite forms under such conditions depends on the calcium or potassium content of the rocks in which deposition occurs. If both are present, both minerals form, as in the Colorado Plateau region; if potassium is not present, only tyuyamunite forms, as at Tyuya Muyun, a hill in Turkestan, U.S.S.R., from which the name is derived, where it occurs in cavities and caverns in limestone ($CaCO_3$). Tyuyamunite crystallizes in orthorhombic system as scaly or radiating aggregates. Properties vary according to water content, which varies with humidity. Hardness is about 2, with specific gravity between 3.3 and 4.4. Surfaces created by cleavage show pearly luster. Crystals may be translucent to opaque. Tyuyamunite is also found as a replacement mineral in petrified wood.

AMBLYGONITE, $(Li,Na)Al(PO_4)(F,OH)$ (lithium-sodium aluminum fluophosphate), forms directly by precipitation in late stages of granite formation in pegmatites, associated with other lithium minerals and apatite. Also occurs in high-temperature tin veins with cassiterite and topaz. It is rich in Li; **natromontebrasite** is Na-rich variety; compositions in-between are called **montebrasite**. Amblygonite crystallizes in triclinic system, is brittle, and has 3 well-developed cleavage directions. Normally white, it may be pale green, pink, or blue. Hardness is 5.5-6, specific gravity 3-3.1. Luster is glassy or pearly.

TRIPHYLITE, $LiFePO_4$ (lithium iron phosphate), like amblygonite, is precipitated directly in granite pegmatites in many areas. If it contains manganese instead of some iron, it is called lithiophilite. It crystallizes in orthorhombic system, occurring as massive aggregates or less commonly as prismatic crystals. It is bluish or greenish gray, yellow, brown, or salmon; surfaces are commonly altered to black. Hardness is 4-5, specific gravity 3.3-3.6, varying with Mn content.

Carnotite
Montrose Co., Colo.

Carnotite
Naturita, Colo.

Tyuyamunite
Valencia Co., N. Mex.

Pink Amblygonite
Keystone, S. Dak.

Amblygonite
Pala, Calif.

Tyuyamunite
Fergana, U.S.S.R.

Triphylite
New Hampshire

Triphylite
Peru, Maine

PYROMORPHITE, $Pb_5(PO_4,AsO_4)_3Cl$ (lead chlorophosphate-arsenate), like many phosphates, occurs in weathered zones of lead-bearing veins associated with other phosphates, carbonates, and sulfates. It is formed by oxidation of the ore minerals, gangue, and wall rock by water. As with many alteration products, it is a pseudomorph—that is, its crystals assume the shape of the minerals, galena and cerussite, it replaces. In pyromorphite, phosphate is dominant over arsenate; in the related **mimetite,** arsenate is dominant. Both minerals are found in many lead-producing areas of world.

Pyromorphite crystallizes in hexagonal system, occurring as prismatic or tabular crystals (generally translucent), globular aggregates, or small grains. It may be green, yellow, brown, orange, red, white, or colorless, depending on composition; luster is resinous, streak white. It has no well-developed cleavage, and fracture is irregular. Hardness is 3.5-4, specific gravity 7.0-7.3.

VANADINITE, $Pb_5(VO_4)_3Cl$ (lead chlorovanadate), like pyromorphite, is a secondary mineral formed by alteration of lead ore by water. The chemical and probably structural similarity illustrates the close relationship of phosphates, arsenates, and vanadates. Natural specimens of vanadinite, in fact, commonly contain appreciable phosphate and arsenate substituting for vanadate as well as calcium substituting for lead. No primary vanadium minerals are present; instead, the vanadium (V) is distributed in very small concentrations in solid solution in the primary minerals. The action of near-surface waters serves to concentrate the vanadium by deposition of the vanadates (VO_4). This concentration is responsible for the occurrence of commercially valuable vanadium deposits. Vanadinite crystallizes in hexagonal system, occurring as prismatic crystals (often skeletal) and as needlelike or globular aggregates. It may be red, orange, yellow or brown, with somewhat resinous or adamantine luster. It has conchoidal or uneven fracture, no cleavage. Hardness is 2.75-3, specific gravity 6.5-7.1. Occurs notably in western U.S., Mexico, Argentina, U.S.S.R., Austria, Scotland, Congo, and northern Africa.

TORBERNITE, $Cu(UO_2)_2(PO_4)_2 \cdot nH_2O$ (hydrous copper uranium phosphate), is a secondary mineral formed by alteration of uraninite. It occurs intergrown with autunite notably in Katanga Prov. of Congo; Alps; Cornwall region in England; Flinders Range in South Australia; and U.S. in uranium areas of Southwest and a few pegmatite locations in Appalachians. It crystallizes in tetragonal system, occurring as tabular crystals and as parallel or micaceous aggregates. It is glassy in luster and generally vivid green. Hardness is 2-2.5, specific gravity 3.22. Cleavage is perfect in one direction. Water content varies.

AUTUNITE, $Ca(UO_2)_2(PO_4)_2 \cdot nH_2O$ (hydrous calcium uranium phosphate), is like torbernite in structure, chemistry, and method of formation. The two occur together, but show little solid solution. Autunite is commoner in pegmatites, as in Smoky Mts., N.C. It occurs as thin, tabular crystals and scaly aggregates or crusts. It is yellow or greenish yellow, glassy in luster, and transparent in thin sheets. It is strongly fluorescent under ultraviolet light. Cleavage is perfect in one direction. Hardness is 2-2.5, specific gravity 3.1-3.2.

Pyromorphite
Panamint Mts., Calif.

PYROMORPHITE CRYSTALS
(Idealized)

VANADINITE CRYSTALS
(Idealized)

Torbernite
Old Gunnis Lake Mine,
Cornwall, England

Vanadinite
Apache Mine, Globe, Ariz.

Autunite
North Carolina

Vanadinite
Shafter, Tex.

Autunite
Grafton Center, N.H.

CHROMATES

As is true of transition metals in general, chromium, Cr, combines chemically with oxygen, O, in several oxidation states. Crocoite and tarapacaite are typical of commoner chromates, with Cr in the $+6$ oxidation state. Lopezite is less common, with Cr in the $+5$ state. The chromate ion, $(CrO_4)^{-2}$, is a tetrahedron, like the sulfate ion $(SO_4)^{-2}$. The net negative charges on the chromate ions are balanced by metal ions packed among the tetrahedra and holding the structure together. The size of these ions and the way they can be packed with the tetrahedra determine the structure. Chromates are a source of chromium used in electroplating steel, as in automobile bumpers and trim, and for alloying with iron to form stainless steels. Chromite, $FeCr_2O_4$ (p. 140), is the major source.

CROCOITE, $PbCrO_4$ (lead chromate), is a popular collectors' item because of its striking orange, red, or yellow color and the occurrence of good prismatic crystals with smooth, brilliant faces. It is a secondary mineral, formed by alteration of lead ore and associated with lead carbonate and tungstates, vanadates, and molybdates. Notable occurrences are in Brazil, the Philippines, Tasmania, and Rhodesia. In the U.S., crocoite is found with tungstates in the mines of Inyo and Riverside counties, Calif., and in Maricopa and Pinal counties, Ariz. Relationships between composition and structure are complex. The high specific gravity and insolubility in water distinguish crocoite from other orange chromates.

Crocoite crystallizes in monoclinic system, occurring as well-developed prismatic crystals or in massive or granular form. It has one good cleavage plane. Hardness is 2.5-3, specific gravity about 6. Luster is vitreous to adamantine. Streak is orange. Good crystals may be transparent. It is very stable in air and insoluble in water.

TARAPACAITE, K_2CrO_4 (potassium chromate), is water-soluble and is found with lopezite in Chilean nitrate deposits in Atacama, Tarapaca, and Antofagasta provinces. Occurrence of lopezite and tarapacaite together is interesting in view of their different chromium oxidation states and depends on the chemistry of the water from which precipitation occurred. Tarapacaite crystallizes in orthorhombic system, occurring as transparent, yellow, tabular crystals. It is soft and brittle, with specific gravity of 2.74.

LOPEZITE, $K_2Cr_2O_7$ (potassium dichromate), is found as small aggregates in Chilean nitrate deposits, with tarapacaite. It is interesting because of the reduced state of the chromium. The compound has been well-studied in the form of artificial crystals, but its occurrence as a mineral is limited by its solubility in water. Synthetic material is triclinic, with well-formed crystals exhibiting 3 good cleavage directions. Hardness is 2.5, specific gravity 2.7. Color is orange red; crystals are transparent if well developed.

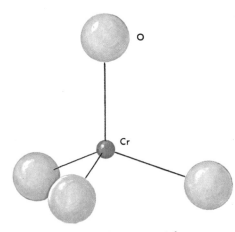

O

Cr

CHROMATE ION $(CrO_4)^{-2}$

CROCOITE CRYSTAL
(Idealized)

Crocoite Crystals
Dundas, Tasmania

Crocoite
Dundas, Tasmania

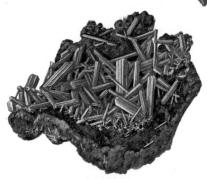

Crocoite
Dundas, Tasmania

Crocoite
Berezov, U.S.S.R.

TUNGSTATES AND MOLYBDATES

The chemistry of the tungstate ion, $(WO_4)^{-2}$, and the molybdate ion $(MoO_4)^{-2}$, is so similar that tungstates and molybdates are always associated, with extensive solid solution, as in the wulfenite group (p. 256). The anions are distorted tetrahedra, with the tungsten atom, W, or the molybdenum atom, Mo, bonded to four oxygen atoms by covalent bonds. Tungstates are particularly important to industry; tungsten metal, alloyed with iron, produces a very strong, corrosion-resistant steel. Tungsten metal is used almost exclusively for making filaments for light bulbs, vacuum tubes, and x-ray tubes. Because tungsten oxidizes readily and burns at high temperature in air, such bulbs and tubes must be evacuated. With development of nuclear reactors for generating electricity, tungsten, molybdenum, and tantalum, all refractory metals, have become important. They are to some extent suitable as cladding for fuel rods containing mixtures of uranium and plutonium oxides because they are non-reactive, have high melting points, and are good conductors of heat.

WOLFRAMITE GROUP is a solid-solution series between huebnerite, $MnWO_4$, and ferberite, $FeWO_4$, with Mn and Fe in the $+2$ oxidation state. Nearly pure end members are rare, and the name wolframite is applied to intermediate compositions. Minerals of the wolframite group are the major ores of tungsten.

WOLFRAMITE, $(Fe,Mn) WO_4$ (iron-manganese tungstate), is a series of high-temperature primary minerals found in sulfide veins and in pegmatites. It occurs in many localities, the largest deposits being in the Nanling mountain range of southern China. Other commercial deposits are found in western U.S.; Cornwall, England; Malay Peninsula; New South Wales and Queensland, Australia; Portugal; Burma; and Bolivia. The wolframite series offers a good example of the variation of properties with composition. The specific gravity, for example, of natural huebnerite is 7.12, but increases uniformly with increasing iron content to 7.51 for natural ferberite. Structural, optical, and physical properties all vary uniformly with composition in the same way.

Wolframite crystallizes in monoclinic (nearly orthorhombic) system, occurring as well-developed prismatic or tabular crystals with one cleavage plane, groups of bladed crystals, massive granular groups, or intergrowths of needlelike crystals. It is brittle, with uneven fracture and hardness of 4-4.5. Luster is submetallic or adamantine. Huebnerite is transparent, iron-rich samples opaque. Color varies with Mn:Fe ratio; it may be brown, yellow, gray, or black, commonly with color banding. Streak is also variable. Ferberite is weakly magnetic. Wolframite minerals are very stable in air.

Wolframite
Bohemia Region,
Czechoslovakia

Huebnerite
Bonita Mt.,
Silverton, Colo.

Huebnerite
Tupiza, Bolivia

WOLFRAMITE, FERBERITE

In Schist with
Quartz

Massive Black
Ore

Massive with
Quartz

WOLFRAMITE, HUEBNERITE

In Quartz
San Juan, Colo.

With Quartz
Missouri

With Quartz
Pima Co., Ariz.

SCHEELITE GROUP is an incomplete solid-solution series between scheelite, $CaWO_4$, and powellite, $CaMoO_4$. The minerals toward the scheelite end of the series, $Ca(WO_4,MoO_4)$, are commoner and are generally called scheelite. The group's structure (facing page) features $(WO_4)^{-2}$ or $(MoO_4)^{-2}$ ions in the form of flattened tetrahedra, with Ca^{+2} ions located between them.

SCHEELITE, $Ca(WO_4,MoO_4)$ (calcium tungstate-molybdate), is a high-temperature primary mineral series, like wolframite. It is found in contact metamorphic deposits, in hydrothermal veins with quartz near granite bodies, and in pegmatites. Scheelite is very widespread and, where abundant, is an important ore of tungsten. Some of the more important deposits are found in northeastern Brazil; near Mill City, Nev.; and in Inyo, San Bernardino, and Kern counties, Calif. Substitution of $(MoO_4)^{-2}$ for $(WO_4)^{-2}$ has been measured in minerals up to a Mo:W ratio of 1:1.38. Some Mg^{+2} substitutes for Ca^{+2}. Substitution of $(WO_4)^{-2}$ for $(MoO_4)^{-2}$ in powellite has been measured up to an Mo:W ratio of approximately 9:1.

Scheelite and powellite crystallize in tetragonal system, occurring as pyramidal crystals, with one good cleavage plane, and as massive, granular, or columnar aggregates. Hardness is 3.5-5, specific gravity 4.2-6.1, depending on composition. Color may be white, yellow, brown, blue, gray, or green. Luster is vitreous, and single crystals are transparent. Both species fluoresce under ultraviolet light.

WULFENITE GROUP, in contrast with scheelite and wolframite groups, consists of secondary minerals, formed by alteration of lead ores with some Mo or W. The wulfenite minerals have the same structure as those of the scheelite group.

WULFENITE, $Pb(MoO_4,WO_4)$, is a solid-solution series between wulfenite, $Pb(MoO_4)$, and stolzite, $Pb(WO_4)$, though complete solid solution has not been shown by natural specimens. It is found in many localities associated with lead ores, from which it was formed by action of near-surface water. Many specimens have Ca^{+2} substituted for Pb^{+2} in large amounts. Unlike other lead compounds, the color is highly variable, with shades of yellow, gray, green, brown, orange, and red. The variation in color results from the variation in composition, not only in terms of the Mo/W ratio but also in terms of substitution of other ions for Pb^{+2}.

Wulfenite crystallizes in tetragonal system, occurring as pyramidal or tabular crystals or as massive or granular aggregates. Hardness is 2.5-3, specific gravity 6.5-8.4 (depending on composition). Single crystals are transparent, with resinous luster.

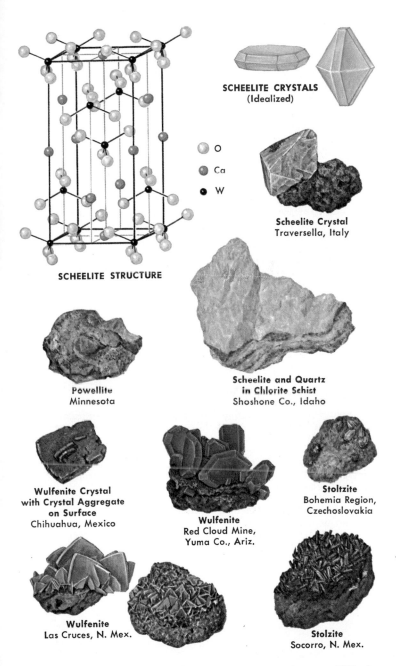

SCHEELITE CRYSTALS
(Idealized)

O
Ca
W

SCHEELITE STRUCTURE

Scheelite Crystal
Traversella, Italy

Powellite
Minnesota

**Scheelite and Quartz
in Chlorite Schist**
Shoshone Co., Idaho

**Wulfenite Crystal
with Crystal Aggregate
on Surface**
Chihuahua, Mexico

Wulfenite
Red Cloud Mine,
Yuma Co., Ariz.

Stoltzite
Bohemia Region,
Czechoslovakia

Wulfenite
Las Cruces, N. Mex.

Stolzite
Socorro, N. Mex.

257

COLLECTING MINERALS

A mineral collection serves several useful purposes for the amateur mineral enthusiast or the professional geologist. Actual specimens enable a person to become familiar with the appearance and physical characteristics of minerals. Several specimens of each mineral are desirable because the color, crystal form, and other characteristics of a mineral may vary to some extent from one specimen to another. In building a collection, one soon becomes adept at mineral identification. The collection then becomes a valuable reference set for comparison with new minerals as they are obtained. The more the collector studies the minerals themselves, the more he will appreciate the beauty and complexity of the mineral kingdom. But perhaps the greatest satisfaction in mineral collecting lies in the discovery of specimens.

Obtaining a large collection is a difficult task, requiring much travel, time, and money. Usually a person cannot collect specimens of all the important minerals by himself because mineral species are not evenly distributed over the earth's surface. But anyone who is fortunate enough to take frequent trips to new areas can find a surprising variety of minerals.

Probably a large number of different minerals can be found near where you live. The bedrock in an area is usually quite uniform and the minerals in it are therefore limited in type. But minerals may have been introduced from other areas. Valleys of the larger streams are often good places for mineral searches because the stream deposits commonly contain minerals carried from far upstream where the bedrock may be different.

Minerals may also have been brought in by glaciers. These ice sheets, which covered extensive areas of the North American continent a few thousand years ago, scoured the complex igneous and metamorphic areas to the north and transported tremendous amounts of minerals to the south. The northern portion of the United States and most of Europe are strewn with glacial debris, known as drift. A search of gravel deposits, particularly in stream beds, will yield a large variety of common rock-forming minerals and even some of the less common ones.

If there is a monument maker near your home, a visit may prove profitable. Much scrap material removed from granite and marble slabs is perfectly suitable for mineral collectors, and the variety of stones used in the manufacture of monuments is large enough to be useful.

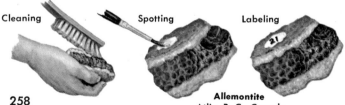

Cleaning   Spotting   Labeling

Allemontite
Atlin, B. C., Canada

GEOLOGIC AND TOPOGRAPHIC MAPS are a big help to the mineral collector. From them he can determine the terrain and rock types of a given area and the collecting possibilities there. Mines, quarries, and similar features of interest are commonly shown.

PURCHASING SPECIMENS from firms that specialize in supplying minerals, rocks, fossils, and other materials to schools, collectors, and professional people is one way of filling gaps in a personal collection. Several of these firms employ collectors who gather specimens from all over the world. These collectors can supply nearly any mineral that a person might want. The firms advertise in mineral periodicals. These same periodicals, with a large circulation among amateur mineralogists (more popularly known as "rockhounds"), also accept classified advertisements from people who want to buy, sell, or trade minerals.

TOOLS AND EQUIPMENT are important to the mineral collector. A short-handled sledge is useful to separate minerals from enclosing rock and to reduce specimens to a size and shape that can be conveniently carried and stored. A smaller hammer, preferably a standard geology hammer with either a pick or chisel head, is a versatile tool for all but the heaviest work. A pair of machinist's goggles should be used to protect the eyes from flying rock chips. Several chisels, wide for splitting rock and narrow for digging into deep pockets, are essential. For separating small fragments of valuable specimens, a dissecting needle and a pair of forceps come in handy.

A small vial of glass or preferably plastic with a screw cap or cork is a convenient place to carry small crystals that might be lost in a large container. For identification, a small hand lens of about 10 power is useful; if it has an attached chain or string it can be worn around the neck. A large canvas knapsack is ideal for carrying specimens when a large number are to be collected.

These tools and a wide assortment of holsters and tool bags plus other equipment are available from firms specializing in mineral-collecting supplies. A collector can also improvise to a considerable extent and devise his own tools and equipment.

Other items that a collector may wish to take with him into the field are: a carpenter's bar for prying rocks apart, newspapers for wrapping specimens, a notebook and pencil, heavy gloves, and possibly a compass. For digging in loose dirt or gravel, a gardener's trowel and hand cultivator may be useful.

Chisels

Goggles

Geology Hammer

Hand Lens

Knapsack

CATALOGING MINERALS is not difficult, but requires a carefully worked out system. The specimens can be numbered in the order in which they are collected. This is the most practical method because it means that acquisition of new specimens does not necessitate a reorganization of the collection. The record-keeping problem can be solved with a carefully kept notebook or filing-card system. The minerals listed in numerical order should include name, chemical formula, crystallographic data, locality from which the mineral came and other information.

The minerals can be cross-indexed in any manner the collector finds convenient. They can be listed by chemical group, by locality, by rock type and so forth. If the collector wants to find a sulfate, for example, he will look under sulfates to get the numbers of the specimens he has. A quick reference to the master list, in numerical order, will provide him with data on any specimen.

It is probably safe to say that the majority of mineral collections are found in cardboard boxes and various drawers, not in any definite order and subject to damage by abrasion and chipping. This is unfortunate, for a good mineral collection is a valuable possession. Inexpensive storage boxes are available in various sizes. As the collection becomes larger and more valuable, it is desirable to acquire special cabinets. Though cabinets are relatively expensive, the convenience with which the minerals can be reached and the protection provided can be well worth the investment.

A SERIOUS INTEREST IN MINERALS is the natural outgrowth of collecting. Like stamp or coin collecting, mineral collecting can become tiresome after a time unless new ideas are encountered and new horizons opened. Just as stamp and coin collecting can lead to an interest in geography and history, so mineral collecting can lead the interested and inquisitive person into the broader fields of geology and chemistry. This progression should be the proper outcome; collecting for its own sake adds nothing to a person's understanding of the world about him.

This book provides basic information about common minerals and their characteristics. The first 69 pages give only a glimpse at the fascinating fields of mineralogy, geology, and chemistry. Some of the references on pages 262-278 will lead the reader to further study of the origin, characteristics, and behavior of minerals.

Altaite
Organ Mts., Dona Ana Co.,
New Mexico

PREPARING MINERAL SPECIMENS is almost as important as finding them in the first place. The size and shape of the mineral collected is normally dictated by the mineral itself. Single crystals or aggregates of crystals occur in a wide range of sizes. Try not to break crystals or separate groups of crystals. Preserving the way in which a mineral occurs in the field will result in the most informative and often the most beautiful specimen.

Many minerals are found as minor components of a massive rock from which they cannot be separated easily. Granular igneous rocks are normally dense, and the feldspar in them is intimately interlocked with quartz and other minerals. In this case it is far better to break off a rock specimen that contains all the essential components than to attempt to separate the component minerals. The entire rock specimen is more meaningful than the separated grains.

Most rocks are relatively fine-grained, and a cubic piece approximately 2 inches on a side is large enough to show the mineralogy and texture. Coarser rocks, such as pegmatites, may have individual crystals larger than a 2-inch cube. These rock samples must therefore be larger to show the complete mineralogy. Single crystals of the individual minerals may be separated from coarse rocks, of course, but samples of the rock itself should also be collected.

IDENTIFICATION should be done, if possible, as the specimen is collected. The locality of the find, the rock type from which the specimen came, and other pertinent geological information should be recorded in a field notebook of the type used by surveyors. The specimen should be numbered, and the number entered in the notebook. A piece of adhesive tape wrapped completely around the specimen serves as a handy label. An even more convenient method of keeping track of specimens is to put each in a sample bag with an attached tag on which field data and the sample number can be written.

PACKING minerals to avoid breakage is extremely important. Wrap each specimen in several layers of newspaper. For very delicate specimens, it may be advisable to use tissue paper or even cotton for protection.

CLEANING is the first step to take, when the mineral is brought home; and, if necessary, to place permanent identification markings on it. The most satisfactory way of doing this is to paint a small spot (about 1/4-inch across) on the specimen, using a good quality white gloss enamel and a small artist's brush. Care should be taken to place the spot of paint where it will not obscure any important feature of the mineral. After the paint is dry, identification numbers can be written on the spot in India ink with a small pen. The numbers should then be entered in a field notebook and cross-indexed in the cataloging system.

LEARNING MORE ABOUT MINERALS

This book was written for the amateur mineralogist and collector to aid in filling the gap between popular works on minerals, many of which are readily available at public libraries, bookstores, and newsstands, and textbooks and reference works, intended for the college student and professional mineralogist. For this reason, only passing reference is made here to the multitude of other books written for the popular market. Rather the intent is to inform the reader of some of the available reference works and textbooks that will aid the serious amateur to proceed beyond the collecting stage to a knowledge of the science of mineralogy, and its relationships with chemistry, physics, and geology. Many of the works mentioned here are highly technical, indeed even incomprehensible, until the vocabulary has been learned. For this reason, an effort has been made to arrange the references, within categories, from the least technical to the most detailed.

It is not wise to invest in most of the works recommended here until one is familiar with the contents and level of complexity. It is suggested, therefore, that the reader visit the library of a nearby college or university for the purpose of examining these books and others. Building the scientific vocabulary necessary to comprehend detailed texts and treatises is a difficult process, but once it has been acquired, understanding technical literature is relatively easy. Most geology and mineralogy teachers at the college level are quite willing to give advice to the amateur as to the most appropriate reading material.

POPULAR MAGAZINES devoted to rocks and minerals are numerous. They provide the reader with articles dealing with the characteristics of minerals, occurrences and associated geology, gem cutting and polishing, and many other interesting topics. They also carry abundant advertising for mineral specimens for sale or trade, notices of coming meetings of national and local mineralogical societies and clubs, and announcements of mineral shows and exhibits. These magazines are immensely useful to the collector and mineral enthusiast who wishes to know of and become affiliated with others having similar interests. The following three magazines are mentioned as representative of those available and no endorsement is implied.

ROCKS AND MINERALS
Official Magazine of the Eastern Federation
of Mineralogical and Lapidary Societies
  Published Monthly
  ROCKS AND MINERALS
  Box 29
  Peekskill, New York 10566, U.S.A.

Clausthalite
Tilkerode, Harz Mts., Germany

LAPIDARY JOURNAL
Published Monthly
LAPIDARY JOURNAL, INC.
3564 Kettner Blvd.
San Diego, California 922112

ROCK AND GEM
Published Monthly
Behn-Miller Publishers, Inc.
16001 Ventura Blvd.
Encino, California 91316

MINERALOGY TEXTBOOKS are numerous and varied in terms of usefulness. A visit to a college or university library to examine current texts is advisable before purchase. The books listed here, with general comments, are included because they contain rather thorough compilations of minerals, along with their properties and occurrences, as well as understandable Introductions to the subjects of determinative mineralogy and crystallography.

MANUAL OF MINERALOGY
James Dwight Dana (1813-1895)
Eighteenth Edition, Revised by Cornelius S. Hurlbut, Jr.
John Wiley, New York, 1971

A standard introductory college text, based on the work of one of the great early mineralogists, a professor at Yale University. Good introductory section provides the essentials of classical mineralogy with some discussion of modern techniques. Contains a rather thorough listing of common minerals, with properties and occurrences.

A TEXTBOOK OF MINERALOGY

Edward Salisbury Dana (1849-1935)
Fourth Edition, Revised and Enlarged by W.E. Ford
John Wiley, New York, 1932

The only thorough compilation of recognized mineral species, based on the work of another great early mineralogist, the son of J.D. Dana and also professor at Yale. The first edition appeared in 1898. Unfortunately, the book is badly out-of-date in terms of chemical and structural treatment, though the basic mineralogy is reasonably accurate. Geographic locations are difficult to interpret because of changes in political boundaries and names since 1932. Though out-of-date, it is still in print and is an indispensable reference.

THE SYSTEM OF MINERALOGY

James Dwight Dana and Edward Salisbury Dana
Seventh Edition
John Wiley, New York

Volume I. Elements, Sulfides, Sulfosalts, Oxides. Entirely Rewritten and Greatly Enlarged by Charles Palache, Harry Berman, and Clifford Frondel, 834 pp., 1944

Volume II. Halides, Nitrates, Borates, Carbonates, Sulfates, Phosphates, Arsenates, Tungstates, Molybdates, etc. Entirely rewritten and greatly enlarged by Charles Palache, Harry Berman, and Clifford Frondel, 1124 pp., 1951

Volume III. Silica Minerals. Entirely rewritten and greatly enlarged by Clifford Frondel, 334 pp., 1962

Intended as the definitive treatise on mineralogy, the multivolume rewriting of the System, written originally by James Dwight Dana in 1837 and revised subsequently by both Danas, father and son, was begun by three distinguished faculty members at Yale University. Volumes I and III, including all the non-silicates, remain the definitive works; Volume III, dealing only with the forms of SiO_2, has since been published, but the silicates, which comprise the great majority of rock-forming minerals, have not yet been treated. More thorough than either the MANUAL or the TEXTBOOK, these volumes are nevertheless somewhat out-of-date, though invaluable.

Alabandite
Queen of the West Mine
Summit Co., Colorado

ROCK-FORMING MINERALS

W.A. Deer, R.A. Howie, and J. Zussman
John Wiley, New York

Volume 1. Ortho- and Ring Silicates,
333 pp., 1962

Volume 2. Chain Silicates,
270 pp., 1963

Volume 3. Sheet Silicates,
270 pp., 1962

Volume 4. Framework Silicates,
435 pp., 1963

Volume 5. Non-Silicates,
371 pp., 1962

This outstanding series of volumes by three noted British mineralogists is the definitive work on the properties and geologic importance of the silicates, which occupy four of the five volumes. They are devoted to the major minerals only and, though they cannot be considered as thorough as the DANA volumes, the depth of information—geologic, structural, chemical, and physical—is remarkable. These volumes were written for the professional geologist and mineralogist and, in general, would not be useful to the amateur or beginning student. A good deal of useful information can be gleaned, however, by one familiar with basic geology and the vocabulary of mineralogy.

CRYSTALLOGRAPHY BOOKS can be singularly frustrating to the beginner. One reason for this is the highly specialized terminology that has been developed by crystallographers to describe the external and internal symmetry of crystalline materials. Another reason is, in part, historical. The crystal systems, classes (point groups), and space groups were determined theoretically by mathematicians and physicists in the late 19th Century, prior to the acquisition of direct information regarding the arrangements of atoms (structure) in real crystals. In general, therefore, crystallography is taught as a rather abstract geometric and mathematical subject, without much reference to crystal structures. The professional crystallographer does, of course, relate the crystallography to the structure, but there are few reference materials from which one can learn such relationships. Only a few crystallography books are mentioned here; study of simpler treatments, as in DANA'S MANUAL is advisable before attempting to use those listed.

AN INTRODUCTION TO CRYSTAL CHEMISTRY

R. C. Evans
Cambridge University Press, 2nd Ed., 1966 (Paperback)

Well written, understandable discussion of crystal structures and chemical bonding. Requires some knowledge of basic chemistry.

INTRODUCTION TO CRYSTAL GEOMETRY
Martin J. Buerger
McGraw-Hill, New York, 1971

A condensed version of the noted crystallographer's ELEMENTARY CRYSTALLOGRAPHY. A good place for the persistent enthusiast to begin studying crystallography.

AN INTRODUCTION TO CRYSTALLOGRAPHY
F. C. Phillips
John Wiley, New York, 3rd Ed., 1963

Technical but readable treatment, with many excellent drawings of crystallographic forms. Treatment of space groups is strictly abstract, with no reference to crystal structures. Good for learning external symmetry (systems and classes).

INTERNATIONAL TABLES FOR X-RAY CRYSTALLOGRAPHY
N. F. M. Henry and Kathleen Lonsdale, Editors
Printed for the International Union of Crystallography by The Kynoch Press, Birmingham, England, 1965

> **Volume I.**   Symmetry Groups
> **Volume II.** Mathematical Tables
> **Volume III.** Physical and Chemical Tables

Published for the professional crystallographer and crystal structure analyst, the INTERNATIONAL TABLES are the authoritative references on crystallographic matters. Strictly not for beginners; useful only after a fundamental knowledge of crystallography has been acquired. They are mentioned here only because anyone interested in minerals and crystals should be aware of their existence.

SYMMETRY IN CHEMISTRY
H. H. Jaffe and M. Orchin
John Wiley, New York, 1965 (Paperback)

SYMMETRY IN MOLECULES
J. M. Holles
Chapman and Hall, London, 1972 (Paperback)

Both are intended for chemists, particularly spectroscopists, who are interested in the symmetry of molecules rather than crystals. Both have excellent descriptions of symmetry operations, necessary for understanding crystallography.

CRYSTAL STRUCTURES, the ways in which atoms are arranged to form the three-dimensional periodic arrays called crystals, are illustrated for many of the minerals described in this book. It is not possible, however, to treat structures thoroughly in a book of this kind, though it is hoped that the reader can gain some appreciation of the relationships among chemistry, structure, and properties. The structures of thousands of crystals have been determined; it is probably not an exaggeration to say that a chemistry lesson can be learned from each. Several references concerning structures are mentioned here:

Tiemannite
Guadalcázar, Mexico

Safflorite
Scheeberg, Saxony

MODELS IN STRUCTURAL INORGANIC CHEMISTRY
A. F. Wells
Oxford University Press, 1970 (Paperback)

TEACHING CHEMISTRY WITH MODELS
R. T. Sanderson
Van Nostrand Reinhold, Princeton, N. J. 1962

These books by two of the "old masters" are very useful to anyone with an interest in the structures of molecules and crystals. Detailed explanations of structures and instructions on how to build accurate models are included. One can learn a lot of chemistry from these little books.

CRYSTAL STRUCTURES OF MINERALS
L. Bragg, G. F. Claringbull, and W. H. Taylor
Volume IV of THE CRYSTALLINE STATE
Cornell University Press, Ithaca, N. Y., 1965

The only comprehensive one-volume compilation of crystal structure information. Can be appreciated and used by anyone with basic knowledge of minerals. Not heavily laden with crystallographic notation.

CRYSTAL STRUCTURES
R. W. G. Wyckoff
Interscience Publ., John Wiley, N. Y., 2nd Ed

Volume 1. Elements, Compounds RX, Compounds RX_2. 467 pp., 1963

Volume 2. Inorganic Compounds RX_n, R_nMX_2, R_nMX_3. 588 pp., 1964

Volume 3. Inorganic Compounds $R_x(MX_y)_z$, $R_x(M_nN_p)_y$, Hydrates and Ammoniates. 981 pp., 1965

Volume 4. Miscellaneous Inorganic Compounds, Silicates, and Basic Structural Information. 566 pp., 1968

Volume 5. The Structures of Aliphatic Compounds. 785 pp., 1966

An attempt by the author, an eminent crystallographer, to compile all available information on crystal structures in one reference set. Though the notation is crystallographically rigorous, the drawings and commentary are very informative to anyone who wishes to know the structure of a mineral or other crystalline materials. The amateur would not want to make the sizable investment required to purchase these volumes, but should be aware of them if his interest carries him deeply into the subject.

CRYSTAL CHEMISTRY, the study of why crystals have a given structure and how that structure is related to properties, is a natural outcome of a serious interest in minerals. It is a rather advanced subject, but one that can be studied by an amateur who knows minerals well. A few interesting books are listed here.

AN INTRODUCTION TO CRYSTAL CHEMISTRY
R. C. Evans
Cambridge University Press, 2nd Ed., 1966 (Paperback)

Well written, understandable discussion of crystal structures and chemical bonding. Requires some knowledge of basic chemistry.

CRYSTALLOGRAPHY AND CRYSTAL CHEMISTRY
F. D. Bloss
Holt, Rinehart & Winston, New York, 1971

The crystallography section is comparable to Phillips' CRYSTALLOGRAPHY. The remainder is broader in scope than Evans' CRYSTAL CHEMISTRY, with excellent chapters on X-ray, optical, and other methods of studying minerals. Useful for the very advanced amateur.

STRUCTURAL INORGANIC CHEMISTRY
A. F. Wells
Clarendon Press, Oxford, 2nd Ed., 1950

More technical and detailed than Evans' CRYSTAL CHEMISTRY. For the advanced reader.

THE NATURE OF THE CHEMICAL BOND
Linus Pauling
Cornell University Press, Ithaca, N. Y., 3rd Ed., 1960

The classic by the Nobel Prize-winning chemist who is in large part responsible for development of the modern theories of chemical bonding. It is heavy reading for the beginner but very informative for one with a good knowledge of general chemistry.

INTRODUCTION TO SOLIDS
L. V. Azaroff
McGraw-Hill, N. Y., 1960

More emphasis on properties of solids than on structures. It contains little information about minerals and requires a good foundation in physics and chemistry but may be useful for the advanced reader who is curious about properties.

CLAY MINERALOGY is a rather specialized area of study within the overall subjects of geology and mineralogy, but the unusual properties of clay minerals and the fantastic breadth of application they have found in industry make the subject worth mentioning.

CLAY MINERALOGY
Ralph E. Grim
McGraw-Hill, N. Y., 2nd Ed., 1968

APPLIED CLAY MINERALOGY
Ralph E. Grim
McGraw-Hill, N. Y., 2nd Ed., 1962

These two books by Ralph E. Grim, Research Professor of Geology, Emeritus, at the University of Illinois, with whom this author had the great privilege of studying for the Ph.D. degree, are detailed summaries of the chemistry, structures, geology, properties, and uses of clay minerals. Though written for the professional, their lucidity makes them quite interesting and informative for the serious amateur.

MINERALOGICAL JOURNALS. Though published for the professional scientist, can be immensely useful to the amateur and the student. Some of the more important journals, though by no means all, are listed here.

THE AMERICAN MINERALOGIST
Journal of the Mineralogical Society of America
Published bimonthly. Established 1916.
Office: Mineralogical Society of America
6th Floor, 1707 L Street, N.W.
Washington, D.C. 20036

Though a professional society, affiliated with the Geological Society of America, membership "is open to persons interested in the fields of mineralogy, petrology, crystallography, and allied sciences." The journal publishes articles devoted to discovery, description, structural determinations, and chemical relationships of minerals. In addition, lists of new mineral names, discredited minerals, new data on known minerals, book reviews, obituaries, and notices of meetings of professional societies. The Mineralogical Society of America provides representatives to the American Crystallographic Association, the International Mineralogical Association, the American Geological Institute, and others.

THE CANADIAN MINERALOGIST

Journal of the Mineralogical Association of Canada
Office: Mineralogical Association of Canada
555 Booth Street
Ottawa, Ontario KIA OGI
Canada

The Mineralogical Association of Canada is organized and functions similarly to other national societies. Though special attention is paid to Canadian mineral occurrences and geology, the journal is international in scope.

MINERALOGICAL MAGAZINE

Journal of the Mineralogical Society
Established 1876
Office: The Mineralogical Society
41 Queen's Gate
London S.W. 7
England

The Mineralogical Society is subdivided into a clay minerals group, an applied mineralogy group, and a geochemistry group, all of which publish technical articles in **Mineralogical Magazine,** one of the oldest and most esteemed of journals. The Society is well organized, providing a full range of services to the membership.

ACTA CRYSTALLOGRAPHICA A. Crystal Physics, Diffraction, Theoretical and General Crystallography

ACTA CRYSTALLOGRAPHICA B. Structural Crystallography and Crystal Chemistry

JOURNAL OF APPLIED CRYSTALLOGRAPHY

These three journals are published for the International Union of Crystallography
Available from:
Munksguard International Booksellers and Publishers, Ltd.
Norre Sogade 35, DK-1370
Copenhagen K, Denmark
and
Polycrystal Book Service
P.O. Box 11567
Pittsburgh, Pennsylvania 15238

These three journals, established since 1948, like the **International Tables for X-Ray Crystallography,** are official publications of the International Union of Crystallography, which serves as the archivists and arbiters for the scientific community. Though highly technical, subject matter should be interesting and informative to the serious amateur and student of mineralogy.

ZEITSCHRIFT FUR KRISTALLOGRAPHIE

Akademische Verlagsgesellschaft
Frankfurt am Main
Germany

This highly respected journal, published in German, English, and French, is devoted to structural and crystallographic matters.

THE CHEMICAL ELEMENTS

Throughout this book the chemical elements are generally denoted by the internationally accepted chemical symbol, consisting of one or two letters, e.g., C for carbon and Mg for magnesium. For the reader's understanding and convenience, the following table lists the elements by name, in alphabetical order, with the correct symbol, the atomic number (the number of protons In the nucleus and the number of electrons in the neutral atom), and the atomic weight (the weight in grams of one mole (6.02 x 10²³ atoms, or 6 followed by 23 zeroes). The atomic weights are averages of the weights of the isotopes that make up the elements as they occur in nature. Atomic weights in parentheses are those of the most common isotopes of elements which occur only as unstable isotopes. Prior to 1966, atomic weights were based on the assumption that natural oxygen had an atomic weight of 16.000. In 1966 the weights were redetermined, using the carbon 12 isotope as the standard reference element and so there are some differences in weights reported in different references.

TABLE OF THE EARTH'S CHEMICAL ELEMENTS

Name	Symbol	Atomic No.	Atomic Wt.	Other Names
Actinium	Ac	89	(227)	
Aluminum	Al	13	26.9815	
Americium	Am	95	(243)	
Antimony	Sb	51	121.75	Stibium
Argon	Ar	18	39.948	
Arsenic	As	33	74.9216	
Astatine	At	85	(210)	
Barium	Ba	56	137.34	
Berkelium	Bk	97	(247)	
Beryllium	Be	4	9.0122	
Bismuth	Bi	83	208.980	
Boron	B	5	10.811	
Bromine	Br	35	79.90	
Cadmium	Cd	48	112.40	
Calcium	Ca	20	40.08	
Californium	Cf	98	(251)	
Carbon	C	6	12.01115	
Cerium	Ce	58	140.12	
Cesium	Cs	55	132.905	
Chlorine	Cl	17	35.453	

Name	Symbol	Atomic No.	Atomic Wt.	Other Names
Chromium	Cr	24	51.996	
Cobalt	Co	27	58.9332	
Copper	Cu	29	63.546	Cuprum
Curium	Cm	96	(247)	
Dysprosium	Dy	66	162.50	
Einsteinium	Es	99	(254)	
Erbium	Er	68	167.26	
Europium	Eu	63	151.96	
Fermium	Fm	100	(257)	
Fluorine	F	9	18.9984	
Francium	Fr	87	(223)	
Gadolinium	Gd	64	157.25	
Gallium	Ga	31	69.72	
Germanium	Ge	32	72.59	
Gold	Au	79	196.967	Aurum
Hafnium	Hf	72	178.49	
Helium	He	2	4.0026	
Holmium	Ho	67	164.930	
Hydrogen	H	1	1.00797	
Indium	In	49	114.82	
Iodine	I	53	126.9044	
Iridium	Ir	77	192.2	
Iron	Fe	26	55.847	Ferrum
Krypton	Kr	36	83.80	
Lanthanum	La	57	138.91	
Lawrencium	Lr	103	(257)	
Lead	Pb	82	207.19	Plumbum
Lithium	Li	3	6.939	
Lutetium	Lu	71	174.97	
Magnesium	Mg	12	24.312	
Manganese	Mn	25	54.9380	
Mendelevium	Md	101	(256)	
Mercury	Hg	80	200.59	Hydrargyrum
Molybdenum	Mo	42	95.94	
Neodymium	Nd	60	144.24	
Neon	Ne	10	20.183	
Neptunium	Np	93	(237)	
Nickel	Ni	28	58.71	
Niobium	Nb	41	92.906	Columbium, Cb
Nitrogen	N	7	14.0067	
Nobelium	No	102	(254)	

Name	Symbol	Atomic No.	Atomic Wt.	Other Names
Osmium	Os	76	190.2	
Oxygen	O	8	15.9994	
Palladium	Pd	46	106.4	
Phosphorus	P	15	30.9738	
Platinum	Pt	78	195.09	
Plutonium	Pu	94	(244)	
Polonium	Po	84	(209)	
Potassium	K	19	39.102	Kalium
Praseodymium	Pr	59	140.907	
Promethium	Pm	61	(145)	
Protactinium	Pa	91	(231)	
Radium	Ra	88	(226)	
Radon	Rn	86	(222)	
Rhenium	Re	75	186.2	
Rhodium	Rh	45	102.905	
Rubidium	Rb	37	85.47	
Ruthenium	Ru	44	101.07	
Samarium	Sm	62	150.35	
Scandium	Sc	21	44.956	
Selenium	Se	34	78.96	
Silicon	Si	14	28.086	
Silver	Ag	47	107.868	Argentum
Sodium	Na	11	22.9898	Natrium
Strontium	Sr	38	87.62	
Sulfur	S	16	32.064	
Tantalum	Ta	73	180.948	
Technetium	Tc	43	(97)	
Tellurium	Te	52	127.60	
Terbium	Tb	65	158.924	
Thallium	Tl	81	204.37	
Thorium	Th	90	232.038	
Thulium	Tm	69	168.934	
Tin	Sn	50	118.69	Stannum
Titanium	Ti	22	47.90	
Tungsten	W	74	183.85	Wolfram
Uranium	U	92	238.03	
Vanadium	V	23	50.942	
Xenon	Xe	54	131.30	
Ytterbium	Yb	70	173.04	
Yttrium	Y	39	88.905	
Zinc	Zn	30	65.37	
Zirconium	Zr	40	91.22	

INDEX

274

C   D   E